Electrical and Electronic Measurements

Electrical and Electronic Measurements

Jasper Andrade

www.clanryeinternational.com

Clanrye International,
750 Third Avenue, 9th Floor,
New York, NY 10017, USA

ISBN: 978-1-64726-646-2

Cataloging-in-Publication Data

Electrical and electronic measurements / Jasper Andrade.
p. cm.
Includes bibliographical references and index.
ISBN 978-1-64726-646-2
1. Electric measurements. 2. Electronic measurements.
3. Electromagnetic measurements. I. Andrade, Jasper.
TK275 .E44 2023
621.37--dc23

For information on all Clanrye International publications
visit our website at www.clanryeinternational.com

Contents

Preface

The world is advancing at a fast pace like never before. Therefore, the need is to keep up with the latest developments. This book was an idea that came to fruition when the specialists in the area realized the need to coordinate together and document essential themes in the subject. That's when I was requested to be the editor. Editing this book has been an honour as it brings together diverse authors researching on different streams of the field. The book collates essential materials contributed by veterans in the area which can be utilized by students and researchers alike.

Electrical measurements are the calculations, methods, and devices utilized for measuring electrical quantities. Electrical and electronic measurement as a discipline deals with the measurement methodologies and procedures by utilizing electric and electronic technologies. It also focuses on the implementation, testing, characterization, design, and calibration of measurement systems. The measurement of electrical quantities is used for determining a system's electrical parameters. Physical properties like pressure, force, temperature, and flow can be transformed into electrical signals by using transducers to measure and record these properties. In day-to-day industrial practice, less exact measurements are required, whereas high precision measurements are utilized in experiments for determining speed of light, charge of an electron and more. This book is a valuable compilation of topics, ranging from the basic to the most complex advancements in the field of electrical and electronic measurements. It strives to provide a fair idea about this discipline and to help develop a better understanding of the latest advances within this field. The book is appropriate for students seeking detailed information in this area as well as for experts.

Each chapter is a sole-standing publication that reflects each author´s interpretation. Thus, the book displays a multi-facetted picture of our current understanding of application, resources and aspects of the field. I would like to thank the contributors of this book and my family for their endless support.

Jasper Andrade

1

Introduction

1.1 Measurement and Error

Classification of Measuring Instruments

Electrical measuring instruments are mainly classified as:

Indicating Instruments

These instruments make use of a dial and pointer for showing or indicating the magnitude of unknown quantity. The examples are ammeters, voltmeter etc.

Recording Instruments

These instruments give a continuous record of the given electrical quantity which is being measured over a specific period.

The examples are various types of recorders. In such recording instruments, the readings are recorded by drawing the graph. The pointer of these instruments is provided with a marker i.e. pen or pencil, which moves on graph paper as per the reading. The X-Y plotter is the best example of such an instrument.

Integrating Instruments

These instruments that measure the total quantity of electricity delivered over period of time. For example, a household energy meter registers the number of revolutions made by the disc to give the total energy delivered, with the help of counting mechanism consisting of dials and pointers.

Functional Elements

The measurement system contains three main functional elements. They are:

- Primary sensing element.
- Variable conversion element.
- Data presentation element.

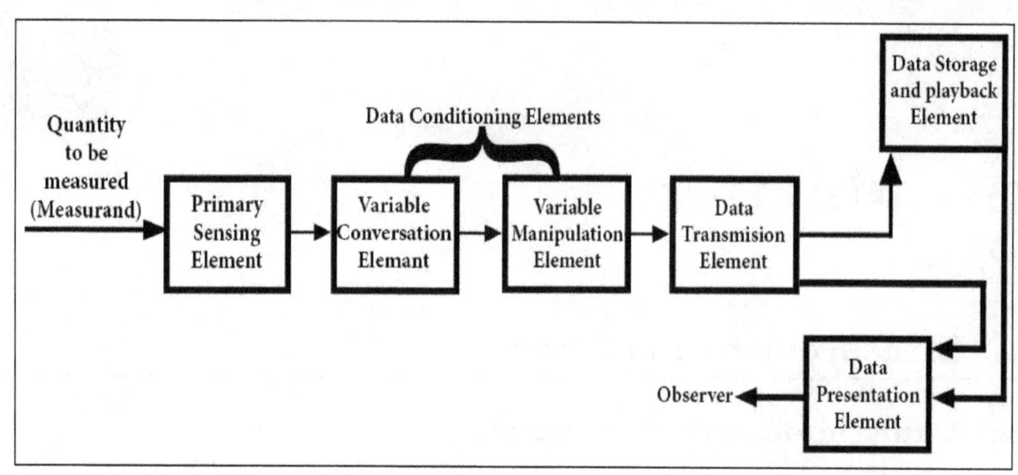

Fundamental elements of an instrument.

Primary Sensing Element: The quantity under measurement makes its first contact with primary sensing element of a measurement system. The primary sensing element is transducers. This transducers converts the measured value into an analogous electrical signal.

Data Conditioning Elements: Data conditioning can be done by two elements. They are variable conversion element and variable manipulation element.

Variable conversion element will convert the output of primary sensing element to some other suitable form to perform the desired function. Example: ADC.

Variable manipulation element is to manipulate the signal given to it preserving the original nature of the signal. Example: Electronic amplifier.

Data Presentation Element: The output or the data of system can be monitored by using visual display devices. These devices may be analog or digital device or even a recording device.

When the element of an instrument is physically separated, it becomes necessary to transmit data from one to another. Therefore transmitting element is used in the measuring system.

1.1.1 Accuracy and Precision and Significant Figures

The performance characteristics of an instrument are mainly divided into two categories:

Static Characteristics

The static characteristics of an instrument are considered for the instruments which are used to measure an unvarying process condition.

The static characteristics of an instrument are:

- Accuracy
- Precision
- Except
- Sensitivity
- Resolution
- Threshold
- Zero Drift
- Reproducibility
- Stability
- Sensitivity drift

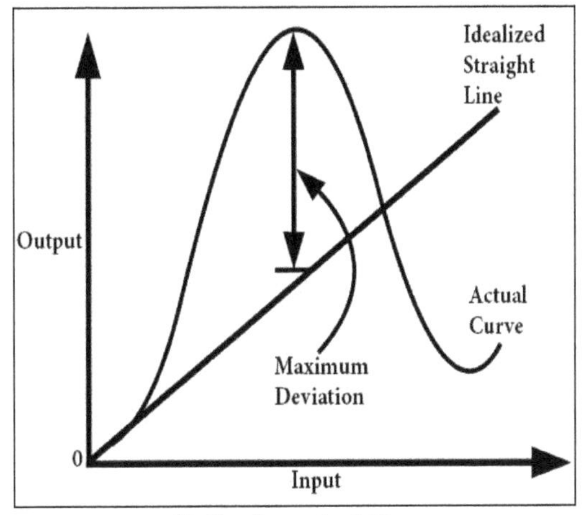

Static characteristics.

Accuracy

It is a measure of the closeness to which an instrument measures the true value of quantity.

Precision

It is the measure of the consistency or repeatability of a series measurements. Although accuracy implies precision, precision does not necessarily imply accuracy. The precision of a given measurement can be given by,

$$\text{Precision} = 1 - \left| \frac{X_i - \bar{X}}{X_i} \right|$$

X_i – Value of ith measurement.

$\bar{X}$ – Average value of n measurements.

Sensitivity

It is a measure of the change in reading of an instrument for a given change in the measured quantity.

$$\text{Sensitivity} = \frac{\text{Infinitestimal change in output}}{\text{Infinitestimal change in input}}$$

Resolution

It is the smallest change in the measured quantity th at will produce a deductible change in the instrument reading.

Error

Error is the deviation from the true value of the measured quantity. Error can be expressed as absolute quantity or as percentage.

Absolute error = $X_e - X_m$

$$\%\ \text{Error} = \left|\frac{X_e - X_m}{X_e}\right|$$

X_e – Expected value.

X_m – Measured value.

Range

The range of an instrument describes the limit of magnitude for which a quantity may be measured. It is normally specified by stating its lower and upper limits.

Span

The span of an instrument is the algebraic difference between the upper and easier limits of the instrument range.

Drift

It is the variation of measured value with time. Perfect re-productibility means that instrument has no drift. There are 3 types of drifts. They are zero drift, span drift and zonal drift.

Real Zone

It is the largest change of input quantity to which there is no output of the instrument. The factor which produces dead zone are hysteresis and backlash in the instrument.

Threshold

The minimum value below which no output change can be detected. This minimum value is called as threshold of the instrument.

Significant Figures

In stating the numerical results of a measurement, only those figures that are meaningful should be recorded. It is important to include only those figures that are justified.

Dynamic Characteristics

The input varies from instant to instant and therefore does the output. The behaviour of the system under such conditions is described by dynamic response.

Dynamic Characteristics of a Measurement Systems.

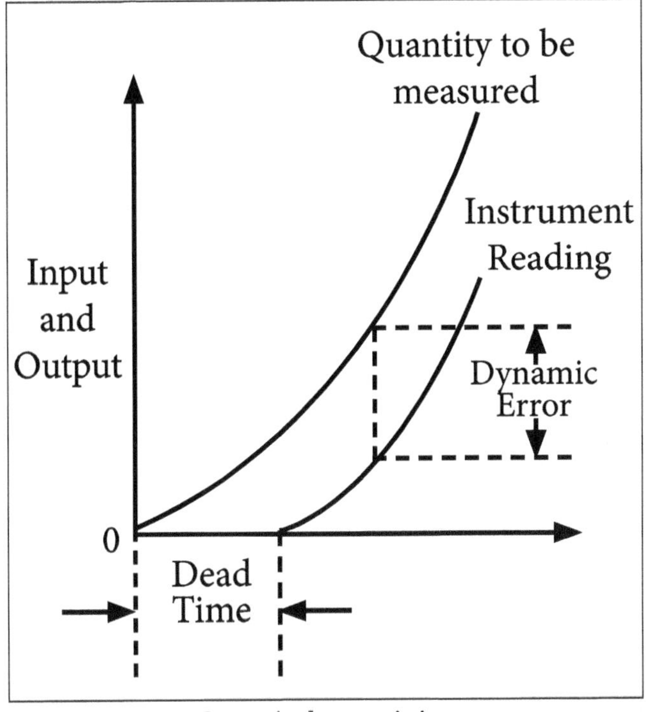

Dynamic characteristics.

Speed of the Response

The speed of the response is defined as the rapidity with which a measurement system responds to the changes in the measured quantity.

Measuring Lag

It refers to changes in measured quantity. The measuring lags are of the following two types: retardation or delay in the response of a measuring system

- Retardation Type Lag: In this type of lag, the response of the measurement system begins immediately after a change in the measured quantity has occurred.
- Time Delay Type Lag: In this case, the response of the measurement system begins only after a dead time after the application of the input.

The measurement lags of this type are very small and are of order of a fraction of second and hence can be ignored.

Fidelity

It is defined as degree to which a measurement system indicates changes in the measured quantity without any dynamic error. It refers to the ability of a system to reproduce the output in the same form as the input.

Dynamic Error

It is the difference between the true value of a quantity changing with time and the value indicated by the measurement system, if no static error is assumed.

The maximum amount by which the pointer moves beyond the steady state is called overshoot.

1.1.2 Types of Errors

Types of Error

The static error is defined as the difference between the true value of this variable and the value indicated by the instruments.

The three types of error are:

1. Gross Error

- The gross errors mainly occur due to lack of experience of a human being.
- These are human mistakes in readings, recordings and calculating results these errors also occur due to incorrect adjustments of instruments. These errors cannot be treated mathematically.
- Some gross errors easily detected with others were very difficulty to detect.

Control Measures to Minimize the Errors:

- Taking great care while taking the reading, recording the reading and calculating the result.
- Without depending on only one readings. At least three or even more readings must be taken and preferably by different persons. The readings must be taken preferably under the condition in which the instrument are switched on and off.

2. Systematic Errors

The systematic errors are mainly resulting due to the short coming of the instruments and the characteristics of the material used in the instrument.

A constant uniform deviation of the operation of an instrument is known of a systematic error.

There are three types of systematic error:

Instrumental Error:

There errors can be mainly due to the following the readings:

- Short comings of instrument: These are because of the mechanical structure of the instruments. For example, Friction in the bearings of various moving parts of the instruments.
- Misuses of instruments: A good instrument if used in abnormal wear gives misleading results, poor initial adjustments, improper, zero setting using leads of high resistance etc.
- Loading effect: Loading effect due to improper way of using the instrument cause the serious errors. The best example of such loading effect error is connecting a well calibrated voltmeter across the two points of high resistance circuit.

Environmental Errors:

These errors are due to the conditions external to the measuring instrument. The various factors resulting in these environmental errors are temperature changes, pressure changes, thermal emf etc.

Steps to minimize these errors:

- Using a proper correction factors and using the information supplied by the manufacture of the instrument.
- The effects of the external fields can be minimized by using the magnetic or electrostatic shields or screens.

Observational Error:

These are the errors introduced by the observer. There are many sources of observational errors such as parallax errors while reading a meter, wrong scale selection, the habits of individual observers etc.

To eliminate such observational errors we should use the instruments with mirrors, knife edged pointers etc.

3. Random Errors

Some errors still result, though the systematic and instrumental errors are reduced (or) at least accounted for. The causes of such errors are unknown and hence the errors are called random error.

The only way to reduce these errors is by increasing the number of observations and using the statistical methods to obtain the best approximation of reading.

Problem

A PMMC ammeter gives reading of 40 mA when connected across two opposite corners of a bridge rectifier, the other two corners of which are connected in series with a capacitor to 100 k, 50 Hz supply. Let us determine the capacitance.

Solution:

Given data:

$$i_{av} = 40 \text{ mA}$$

$$V_m = 100 \text{ k}$$

$$f = 50\text{Hz.}$$

Formula to be used:

$$V_m = \frac{i_{av}}{4C_f}$$

$$\text{W.K.T, } V_m = \frac{i_{av}}{4C_f}$$

$$= C = \frac{i_{av}}{V_m 4_f} = \frac{40\times10^{-3}}{100\times10^{3}\times4\times50}$$

$$= 2\times10^{-9} \text{ F.}$$

1.2 Standards of Measurement and Classification of Standards

Different Standards of an instrument are as follows:

International Standards

The international standards are defined on the basis of international agreement. They represent the units of measurements which are closest to the possible accuracy attainable with present day technological and scientific methods. International standards are checked and evaluated regularly against absolute measurements in terms of the fundamental units. The International Standards are maintained at the International Bureau of Weights and Measures and are not available to the ordinary user of measuring instruments for the purposes of calibration and comparison.

Improvements in the accuracy of absolute measurements have made the international units superfluous and they have been replaced by absolute units. One of the main reasons for adopting an absolute system of units is that new standards for resistance, can be constructed which are sufficiently permanent and do not vary appreciably with time.

Primary Standards

Primary standards are absolute standards of such high accuracy that can be used as the ultimate reference standards. These standards are maintained by national standards laboratories in different parts of the world. The primary standards, which represent the fundamental units and some of the derived electrical and mechanical units, are independently calibrated by absolute measurements at each of the national laboratories. The results of these measurements are compared against each other, leading to a world average figure for the primary standards. Primary standards are not available for use outside the national laboratories. One of the primary standards is the verifications and calibration of secondary standards.

The primary standards are few in number. They must have the highest possible accuracy. Also these standards must have the highest stability. i.e., their values should vary as small as possible over long periods of time even if there are environmental and other changes.

In the recent past, the techniques of establishing primary standards have been drastically refined so that accuracy attainable has a very high level.

Secondary Standards

The secondary standards are the basic reference standards used in industrial measurement laboratories. The responsibility of maintenance and calibration of these standards lies with the particular industry involved. These standards are checked locally against reference standards, available in the area.

Secondary standards are normally sent periodically to the natural standards laboratories for calibration and comparison against primary standards. The secondary standards are sent back to the industry by the national laboratories with a artification regards and their measured values in terms of primary standards.

Working Standards

The working standards are the major tools of a measurement laboratory. These standards are caused to check and calibrate general laboratory instrument for their accuracy and performances. For example, a manufacturer of precision resistances, may use a 'Standard Resistance' which may be a working standard, in the quality control department for checking the names of resistors that are being manufactured. This way be verifies that his measurement set up performs within the limits of accuracy that are specified.

Calibration is defined as the comparison of instrument with a primary or secondary standard of an instrument of known accuracy.

Static Calibration

It is defined as the process by which all the static performance characteristics are obtained in one form or another. Calibration is used to find errors.

Calibration procedure may be classified as:

Primary Calibration

Instrument is calibrated against primary standards. Some of the examples are standard resistor and standard cell. After calibration, that instrument is called as secondary calibration.

Secondary Calibration

Secondary calibration instrument is used for calibrating another instrument of lesser accuracy. It is widely used in laboratory practice as well as in industry.

Example: Standard cell may be used for calibrating a voltmeter or ammeter with a suitable circuit.

Direct Calibration

It is same order of accuracy as primary calibration. Instruments which are calibrated directly are used as secondary calibration instruments.

Example: flow meter.

Indirect Calibration

It is equivalence of two different instruments used for measurement of certain physical quantity.

Consider two flow meters having similarity Reynolds number.

i.e.,

$$\frac{D_1 \rho_1 V_1}{\mu_1} = \frac{D_2 \rho_2 V_2}{\mu_2}$$

Where,

D → Diameter of the pipe

p → Density of the fluid

V → Velocity of fluid stream

Routine Calibration

It is periodically checking the accuracy and proper functioning of the instrument with standards.

Usual steps of routine calibration are:

- Visual inspection of the instrument for the obvious physical defects.
- Checking the instrument for proper installation in accordance with the manufacturer specifications.
- Zero setting of all the indicators.
- Levelling of instruments which needs this precaution.
- Recommended operation tests to detect for major defects.

Significance of calibration:

The calibration of a measuring instruments by introducing an accurately known sample of variables that is to be measured and then observing the system response.

The measuring instrument is checked and adjusted until its scale reads the accurate sample of the variable. This can be achieved only by calibration.

1.2.1 Electrical Standards

The International conference on electrical units in London in 1908 confirmed the absolute system units adopted by the British Association Committee on Electrical Measurement in 1863. This conference decided to specify some material standard which can be produced in isolated laboratories and used as International standards. The desired properties of International standards are that they should have a definite value, be permanent, and be readily set up anywhere in the world, also that their magnitude should be within the range at which the most accurate measurements can be done. The four units – ohm, ampere, volt and watt – established by above specifications were known as International units. The ohm and ampere are primary standards. Definitions of International unit are given below,

International Ohm

The international ohm is the resistance offered to the passage of an unvarying electric current at the temperature of melting ice by a column of mercury of uniform cross-section, 106.300 cm long and having mass of 14.4521 gm (i.e., about 1 sq. mm in cross-section).

International Ampere

The international ampere is the unvarying current which when passed through a solution of silver nitrate in the water deposits at the rate of 0.0001118 gm per second.

The International Volt and Watt

The international volt and watt are defined in terms of International ohm and ampere. As constructing standards, which did not vary appreciably with time, it was difficult and also as, by 1930, it was clear that the absolute ohm and ampere could be determined as accurately as international units. The International committee on Weights and Measures decided in October, 1946 to abandon the international units and choose January 1, 1948 as date for putting new units into effect. The change was made at an appropriate time and the absolute system of electrical units is now in use as the system on which electrical measurements are based.

1.2.2 IEEE Standards

IEEE is the Institute of Electrical and Electronics Engineers which is headquartered in New York city. There are standard procedures, definitions, and nomenclature instead of physical items standards available for comparing secondary standards The IEEE standards control the instrument's front panels, for test and measuring procedures, for electrical hardware and installations in particular situations.

These standards define certain standard test methods for the evaluation of the various electronic components and devices. Some standards are meant to specify test equipment that each manufacturer must adopt, like the arrangement and names of knobs, controls, and functions of common laboratory oscilloscope. If this is not adopted it becomes difficult for the oscilloscope operator to operate different oscilloscopes differently. There are other standards as well that concern the safety of wiring for ships, industrial buildings, power plants, and many more applications.

The standard voltage and current ratings have been specified so that when any component is to be changed, it can be changed without any damage. The standard logic and schematic symbols have been defined for engineering drawing to make understandable by engineers. The most important IEEE standard is the standard hardware for interfacing the laboratory test equipment to computers for monitoring and control purposes.

1.3 Measurement of Resistance and Inductance

Resistance: Measurement of Low Resistance by Kelvin's Double Bridge.

Low Resistance (Kelvin Double Bridge Method):

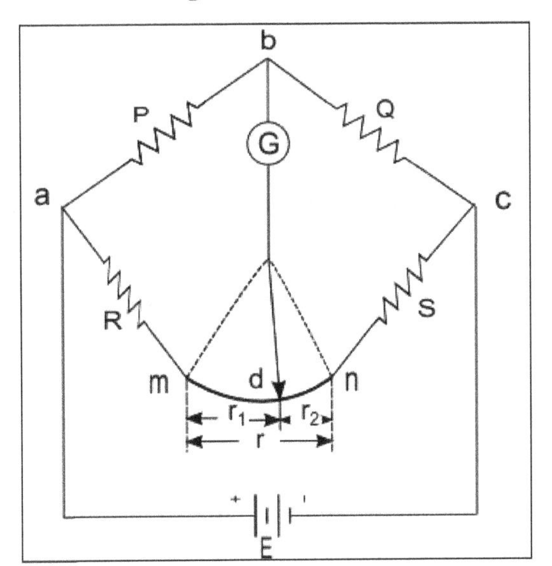

Kelvin bridge method for low resistance measurement.

Kelvin Bridge, a modification of Wheatstone bridge, method increases the accuracy in measurement of low resistance and remove the effect of connecting leads and contact resistance. As shown in figure, r represents the resistance of the lead that connects the unknown resistance R and standard resistance S. The galvanometer is connected at point d that divides resistance r into r_1 and r_2 such that:

$$\frac{r_1}{r_2} = \frac{P}{Q} \qquad ...(1)$$

Using equation (1) we have:

$$\Rightarrow \frac{r_1}{r_1 + r_2} = \frac{P}{P+Q} \Rightarrow r_1 = \frac{P}{P+Q} r \qquad ...(2)$$

$$\Rightarrow \frac{r_1 + r_2}{r_2} = \frac{P+Q}{Q} \Rightarrow r_2 = \frac{Q}{P+Q} r \qquad ...(3)$$

From the figure at balance condition:

$$\frac{R + r_1}{S + r_2} = \frac{P}{Q} \qquad ...(4)$$

$$R + r_1 = \frac{P}{Q}(S + r_2) \qquad ...(5)$$

$$R + \frac{P}{P+Q} r = \frac{P}{Q}\left(S + \frac{Q}{P+Q} r\right) \qquad ...(6)$$

$$R = \frac{P}{Q}.S \qquad ...(7)$$

So, connecting the galvanometer at point d, the resistance of leads does not affect the result. But, the problems with the above method are the method is not practical and difficult to find correct galvanometer null point.

To solve the above problems, two actual resistance unit of correct ratio is connected between point's m and n as shown in figure below which is the original Kelvin Double bridge. The ratio arm of p and q is connected at d to eliminate the effect of connecting leads in between R and S. The value of P, Q, p and q are like that p/q = P/Q.

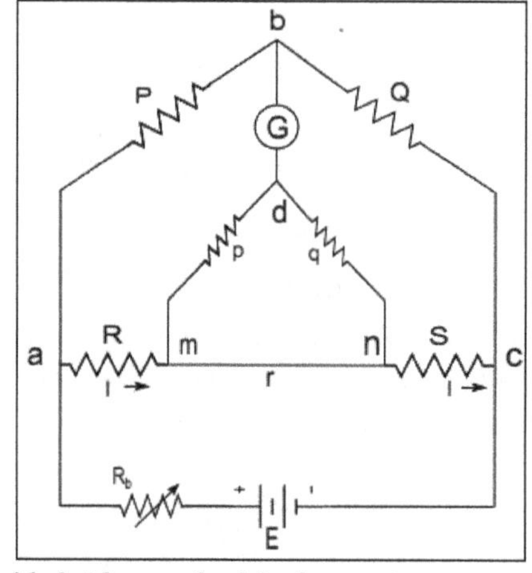

Kelvin Double bridge method for low resistance measurement.

Under balance condition there is no current through galvanometer, which means:

$$E_{ab} = E_{amd}$$

Where,

$$E_{ab} = \frac{P}{P+Q} E_{ac}$$

$$E_{ac} = I\left[R + S + \frac{(p+q)r}{p+q+r}\right]$$

$$E_{amd} = I\left[R + \frac{r}{p+q+r} p\right]$$

At balance condition:

$$E_{ab} = E_{amd} \qquad ...(8)$$

$$\Rightarrow \frac{P}{P+Q} E_{ac} = I\left[R + \frac{r}{p+q+r} p\right] \qquad ...(9)$$

$$\Rightarrow \frac{P}{P+Q} I\left[R + S + \frac{(p+q)r}{p+q+r}\right] = I\left[R + \frac{r}{p+q+r} p\right] \qquad ...(10)$$

$$\Rightarrow R = \frac{P}{Q} S + \frac{qr}{p+q+r}\left[\frac{P}{Q} - \frac{p}{q}\right] \qquad ...(11)$$

$$\Rightarrow R = \frac{P}{Q} S \qquad \text{since, } \frac{p}{q} = \frac{P}{Q} \qquad ...(12)$$

The equation (12) shows that, the resistance of connecting leads has no effect but error may be introduced in ratio arms, i.e. $p/q = P/Q$ may not equal. Thermoelectric effect can be removed by reversing the battery connection and true value of R will be the mean of two readings.

1.3.1 Measurement of Medium Resistance

Medium Resistance

A Wheatstone bridge is used in measurement of medium resistances. It is an accurate and reliable instrument and is extensively used in industry. The wheat stone bridge is an instrument for making comparison measurements and operates upon a null indication principle.

A very high degree of accuracy can be achieved using wheat stone bridge. Accuracy of 0.1% is quite common with a wheat stone bridge.

The figure shows the basic circuit of a wheat stone bridge. It has four resistive arms, consisting of resistance P, Q, R and S together with a source of emf and a null detector usually a galvanometer G or other sensitive current meter. The current through the galvanometer depends on the potential difference between points 'd' and 'b'.

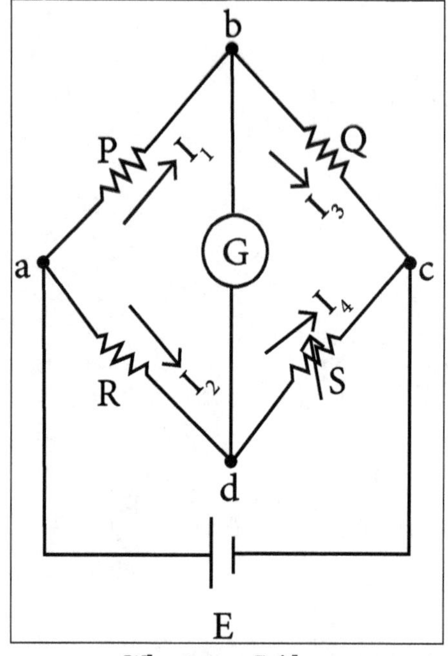

Wheatstone Bridge.

The bridge is said to be balanced when there is no current through the galvanometer or when the potential difference across the galvanometer is zero. This occurs when the voltage from point 'b' to point 'a' equals the voltage from point 'd' to point 'a' or by referring to the other battery terminal, when the voltage from point 'd' by point 'c' equals the voltage from point 'b' to point 'c'.

For bridge balance, we can write:

$$I_1P = i_2R.$$

For the galvanometer current to be zero, the following, conditions also exist:

$$I_1 = I_3 = \frac{E}{P+Q}$$

And,

$$I_2 = I_4 = \frac{}{R \quad Q}$$

Where, E = emf of the battery.

Combining the above equations and simplifying, we obtain:

$$\frac{P}{P+Q} = \frac{R}{R+S}$$

$$\text{From which } QR = PS \qquad \ldots(A)$$

Equation A is the well-known expression for the balance of wheat stone bridge. If three of the resistances are known, then the fourth may be determined from (A), as:

$$R = S\frac{P}{Q}$$

Where, R is the unknown resistance, S is called the standard arm of the bridge and P and Q are called ratio arms.

1.3.2 Measurement of High Resistance

High Resistance

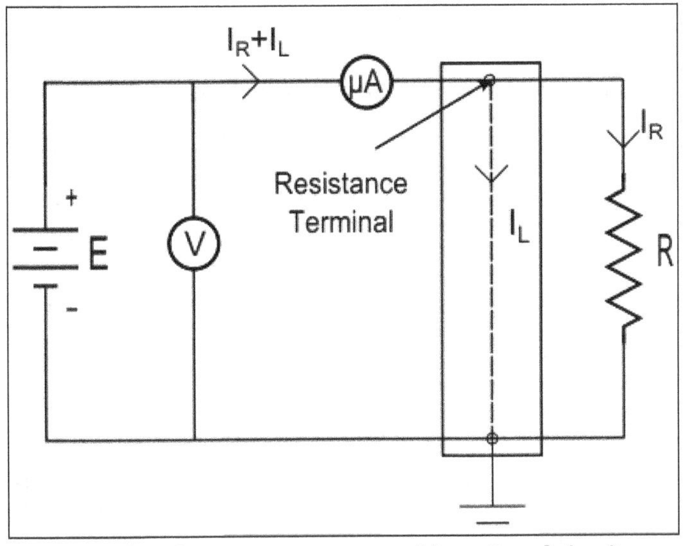

(a)High resistance measurement without Guard circuit.

The main problem in measurement of high resistance is the leakage resistance. To eliminate the errors due to leakage resistance some form of guard circuits are generally used as shown in the figure (a). In figure (b) the high resistance mounted on a piece of insulating material is measured by the ammeter-voltmeter method. The micro-ammeter measures the sum of the current through the resistor (IR) and the current through the leakage path around the resistor (I_L).

Therefore, the measured value of the resistor calculated from the reading of the voltmeter and micro-ammeter will not be the true value of the resistor but will be in error. In figure (b) a guard terminal is added to the resistance terminal block and micro-ammeter is bypassed. Now, micro-ammeter will measure only the current through the resistor R that allows to determine the correct measurement of the resistor.

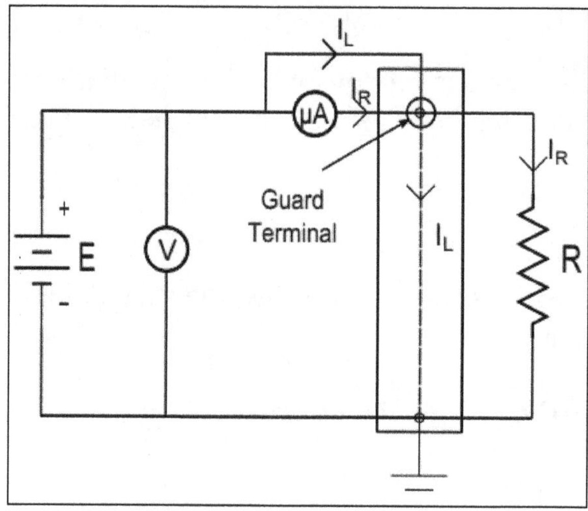

High resistance measurement with Guard circuit.

1.3.3 Measurement of Resistance of Insulating Materials

An insulating material's resistance to the conductive heat flow is measured or rated in terms of its thermal resistance or R-value the higher the R-value, greater the insulating effectiveness. The R-value depends on the type of insulation, its thickness, and its density. When calculating the R-value of the multilayered installation, add the R-values of the individual layers. Installing more insulation in our home increases the R-value and the resistance to heat flow. To determine how much insulation we need for our climate, consult a local insulation contractor.

The effectiveness of an insulation material's resistance to the heat flow also depends on how and where the insulation is installed. For example, insulation that is compressed wont provide its full rated R-value. The overall R-value of a wall or ceiling will be somewhat different from the R-value of the insulation itself because the heat flows more readily through studs, joists, and other building materials, in a phenomenon known as thermal bridging. In addition, insulation that fills building cavities densely enough to reduce airflow can also reduce convective heat loss.

Insulation Resistance Testing

Basically, we are applying a voltage (specifically a highly regulated, stabilized DC voltage) across a dielectric, measuring the amount of current flowing through that dielectric then calculating (using Ohm's Law) a resistance measurement. Let's clarify our use of term "current." We're talking about leakage current. The resistance measurement is in meg ohms. we use this resistance measurement to evaluate insulation integrity.

Current flow through a dielectric may seem somewhat contradictory, but remember, nowhere electrical insulation is perfect. So, some current will flow.

The Purpose of Insulation Resistance Testing

We can use it as:

- A quality control measure at the time a piece of electrical equipment is produced.
- An installation is required to help ensure specifications are met and to verify proper hookup.
- A periodic preventive maintenance task.
- A troubleshooting tool.

Operation of an Insulation Resistance Test

Generally, we connect two leads (positive and negative) across the insulation barrier. A third lead, which connects to a guard terminal, may or may not be available with our tester. If it is available, we may or may not have to use it. This guard terminal acts as a shunt to remove the connected element from the measurement. In other words, it will allow you to be selective in evaluating certain specific components in a large piece of electrical equipment.

Obviously, it's a good idea to have a basic familiarity with the item we're testing. Basically, we should know what is supposed to be insulated from what. The equipment we're testing will determine how you hook up our meghommeter.

After we make our connections, we apply the test voltage for 1 min. (This is a standard industry parameter that allows us to make relatively accurate comparisons of readings from past tests done by other technicians.)

During this interval, the resistance reading should drop or remain relatively steady. Larger insulation systems will show a steady decrease. smaller systems will remain steady because the capacitive and absorption currents drop to zero faster than in larger systems. After 1 min, we should read and record the resistance value.

When performing insulation resistance testing, we must maintain consistency. Why?

Because electrical insulation will exhibit dynamic behavior during the course of our test; whether the dielectric is "good" or "bad." To evaluate a number of test results on the same piece of equipment, we have to conduct the test the same way and under the relatively the same environmental parameters, each and every time.

Our resistance measurement readings will also change with time. This is because the electrical insulation materials exhibit capacitance property and will charge during the course of the test.

This can be somewhat frustrating to a novice. However, it becomes a useful tool to a seasoned technician.

As we gain more skills, we will become familiar with this behavior and be able to make maximum use of it in evaluating your test results. This is one factor that generates the continued popularity of analog testers.

Causes on Insulation Resistance Reading:

Insulation resistance is temperature-sensitive. When temperature increases, insulation resistance decreases, and vice versa. A common rule of thumb is insulation resistance changes by a factor of two for each 10 Deg change. So, to compare new readings with previous ones, we will have to correct our readings to some base temperature. For example, suppose we measured 100 megohms with an insulation temperature of 30 Deg A corrected measurement at 20 Deg would be 200 megohms (100 megohms times two).

Also, "acceptable" values of insulation resistance depend upon the equipment we're testing. Historically, many field electricians use the somewhat arbitrary standard of 1 megohm per kV. The National Electrical Testing Association (NETA) specification Maintenance Testing Specifications for Electrical Power Distribution Equipment and Systems provides much more realistic and useful values.

Remember compare our test readings with others taken on similar equipment. Then, investigate any values below the NETS standard minimums or sudden departures from previous values.

1.3.4 Portable Resistance Testing Set (Megohmmeter) and Measurement of Insulation Resistance when Power is ON

A megger is a universally used testing instrument for the installation tests and it is combined on hand-driven generator and n ohm-meter giving direct readings in ohms and in mega ohms. The megger is the most popular type of insulation and continuity tester. The 500V model of both hand-driven and battery-type instruments are normally used for performing all tests on domestic wiring. In principal this instrument consists of a permanent magnet in the field of which is a voltage and a current coil fixed at an angle to one another with a pointer pivoted at the center of rotation of the coils.

The needle or pointer deflection (The movement of pointer on the scale is known as deflection) is a function of ratio of the current-driven generator it has a permanent magnet and provides test voltage. The open circuit voltage (when the. R is infinity) is generally about 5% of the rated voltage and with zero 'R' the voltage will be practically zero.

There are two scales present on the megger, ohm meter-an ohms scale, normally 0-100Ω or 0-200Ω), for continuity testing; and a mega ohms scale, normally 0-500MΩ) or 0-100MΩ for insulation testing. It is also important in all cases to turn the handle at a steady constant speed of about 160 r.p.m. Because of its extreme simplicity in use, it can .be operated with one-hand. But, the battery operated megger instrument is now much in favour.

In normal condition the pointer of the megger indicate the infinity value or in normal condition the pointer is in stand-still at infinity scale. Megger has two terminals; one is for earth 'E' and other for live 'U. The testing circuit loads are connected to these two terminals. All the wiring test such as, insulation test polarity test, continuity test, short-circuit test are commonly performed by using a megger. We also check the winding insulation of transformers generator and motors by using a megger. It is available in various voltage range such as 500V, 1000V; 2500V. But, now a days electronic insulator meter are also available which works with 6 cells of 1.5 volt each and hence don't require hand operation It can also measure resistance between 0-20 and high resistance of 0.05MΩ to 100MΩ and AC voltage up-to 500 volt:

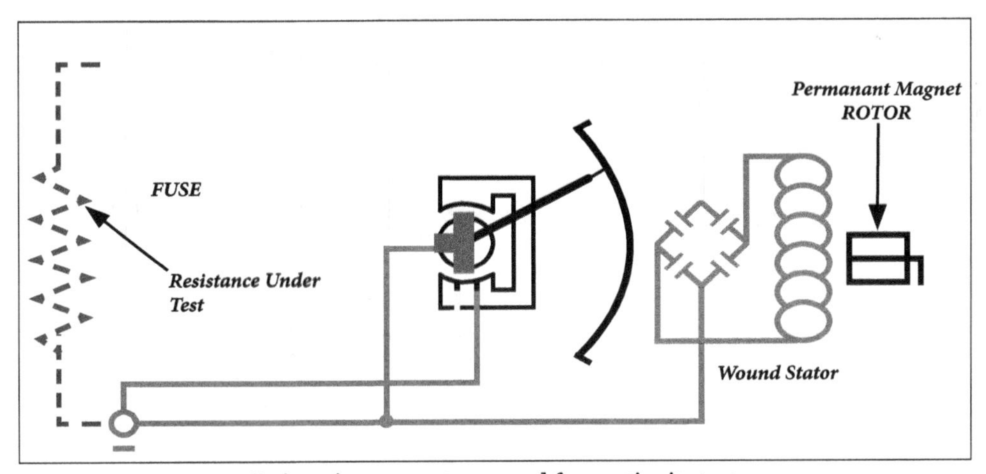

Series 3 instrument arranged for continuity test.

1.3.5 Measurement of Resistance of Earth Connections

The earth resistance should be as minimum as possible to: protect various parts of insulations, high voltage discharge and for stabilizing $3-\Phi$ circuit. The methods of measuring earth resistance:

- Fall of Potential Method.

- Earth Tester.

Fall of Potential Method

As shown in figure a current is passed through earth electrode E to another electrode B. The lines of first electrode current diverge and those of second electrode current converge. As a result the current density is much greater in the vicinity of the electrodes than at a distance from them. The potential distribution between the electrodes is shown in figure (c). It is obvious from the curve that the potential rises in the proximity of electrodes E and B and is constant along the middle section.

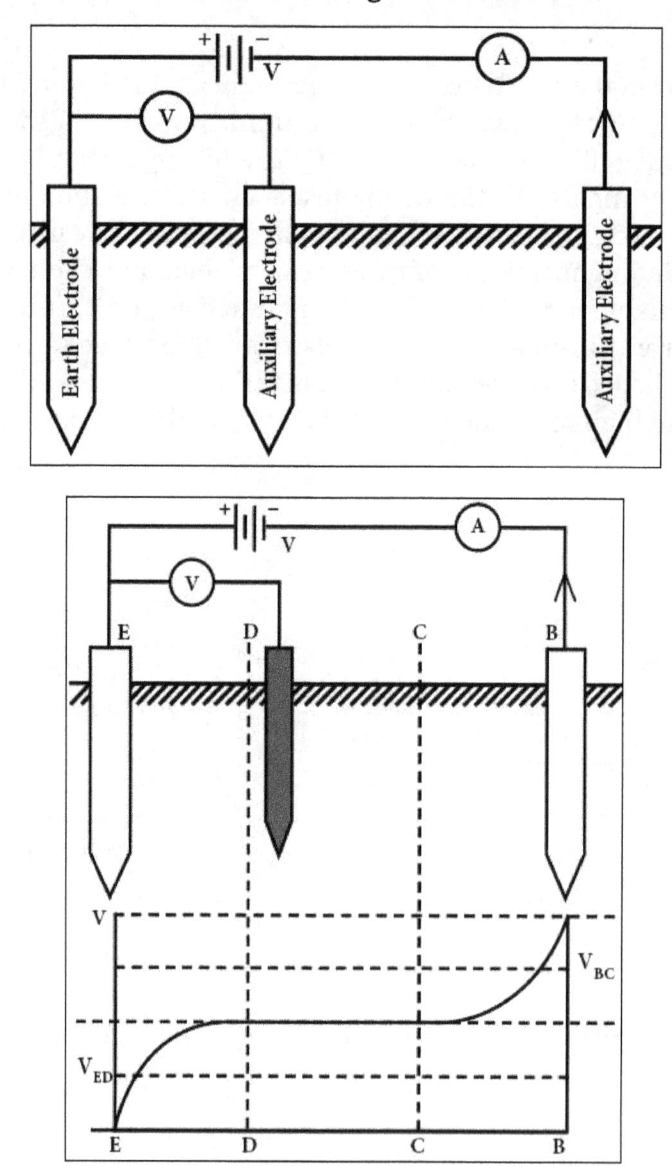

RE = V / I = VEA / I

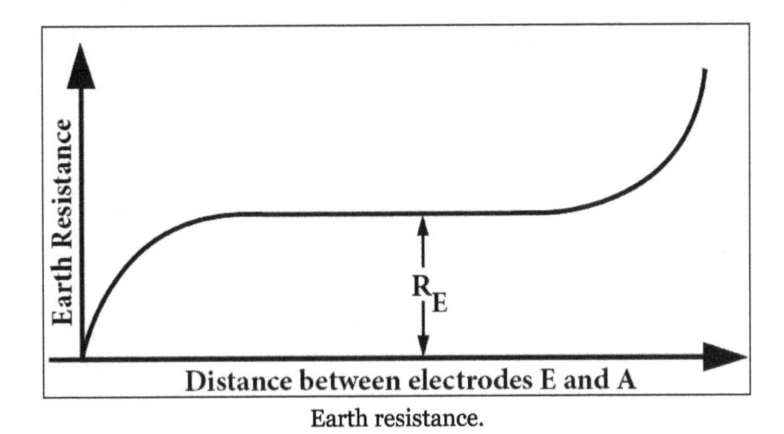

Earth resistance.

Localization of Cable Faults

Fault occurring in cables which are in use on lower distribution voltage. The common faults are: Ground fault (core of the cable to ground) and short circuit fault (core of one cable to that of another cable). The methods for localizing these type of cable faults are:

Murray Loop Test

The connection diagram of this method is shown in figure. The resistances P, Q, R and X forms essentially a Wheat stone bridge.

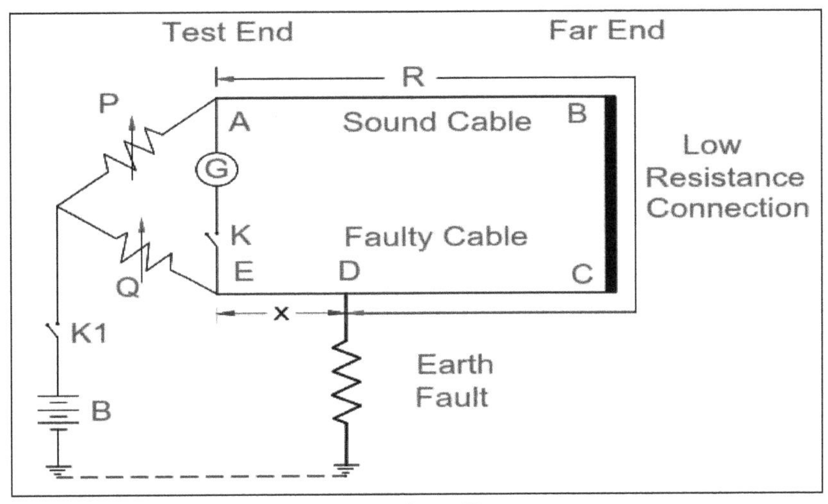

Murray Loop Test.

Under balance condition:

$$\frac{P}{R}=\frac{Q}{R}\Rightarrow\frac{X}{R}=\frac{Q}{P}\Rightarrow\frac{X}{X+R}=\frac{Q}{P+Q} \qquad \text{...(1)}$$

$$X=\frac{Q}{P+Q}(R+X) \qquad \text{...(2)}$$

If l_1 represents the length of the fault from the test end and l is the length of each cable then:

$$l_1 = \frac{Q}{P+Q}.2l \qquad ...(3)$$

Therefore, the position of the fault may be located if the length of the cable is known. In this test it seen that the fault resistance does not alter the balance condition because it enters the battery circuit. But, if the fault resistance is high, this may reduce the sensitivity of the bridge and accurate measurement will be impossible. This high resistance effect may be reduced by applying high dc or ac voltage that may carbonize the insulation of the cable at the point of fault.

Varley Loop Test

The necessary connection diagram for this test is shown in figure. An SPDT switch K is set at position 1 and balance is obtained by varying S.

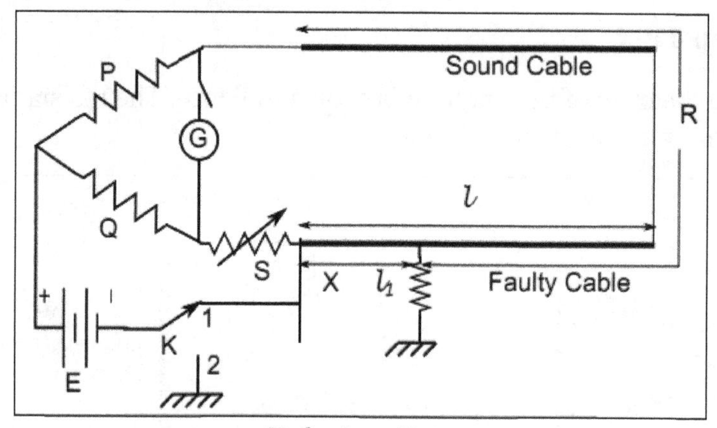

Varley Loop Test.

Let, for first case the value of S is S_1. Under balance condition:

$$\frac{P}{R+X} = \frac{Q}{S_1} \Rightarrow \frac{R+X}{S_1} = \frac{P}{Q} \qquad ...(4)$$

$$R+X = \frac{P}{Q}S_1 \qquad ...(5)$$

The switch K is thrown to position 2 and bridge is rebalanced. Then, the balance condition given for the value of S is S_2:

$$\frac{P}{R} = \frac{Q}{X+S_2} \qquad ...(6)$$

$$\frac{R}{X+S_2}=\frac{P}{Q} \quad ...(7)$$

$$\frac{R+X+S_2}{X+S_2}=\frac{P+Q}{Q} \quad ...(8)$$

$$X=\frac{(R+X)Q-S_2P}{P+Q} \quad ...(9)$$

The value of X is obtained from equation (9) with the help of equation (5). For the cables of same cross section and resistivity, the resistances are proportional to the length. Now, if X = lr and R + X = 2lr then:

$$\frac{X}{R+X}=\frac{l_1}{2l} \quad ...(10)$$

$$l_1=\frac{X}{R+X}2l \quad ...(11)$$

1.4 Inductance: Measurement of Self-inductance

Quality Factor

The concept of quality factor is one that is applicable in many areas of physics and engineering and it is denoted by letter Q and may be referred to as the Q factor.

The Q factor is a dimensionless parameter that indicates energy losses within a resonant element which could be anything from a mechanical pendulum, an element in a mechanical structure, or within an electronic circuit such as a resonant circuit. In particular Q is often used in association with an inductor.

While the Q factor of an element relates the losses, this links directly in to the bandwidth of the resonator with respect to its center frequency. As such the Q or quality factor is particularly important within RF tuned circuits, filters, etc..

The Q indicates energy loss relative to the amount of energy stored within the system. Thus the higher the Q the lower the rate of energy loss and hence oscillations will reduce more slowly, that is they will have a low level of damping and they will ring for longer.

For electronic circuits, energy losses in the circuit are caused by resistance. Although this can occur anywhere within the circuit, the main cause of resistance occurs within an inductor. Accordingly inductor Q is a major factor within resonant circuits.

Quality Factor Definition

The definition of quality factor is often needed to give a more exact understanding of what this quantity actually is.

For electronic circuits, Q is defined as the ratio of the energy stored in the resonator to the energy supplied to it, per cycle, to keep signal amplitude constant, at a frequency where the stored energy is constant with time.

It can also be defined for an inductor as the ratio of its inductive reactance to its resistance at a particular frequency and it is also a measure of its efficiency.

Q Factor Equations

The basic Q or quality factor equation is based upon energy losses within the inductor, circuit or other form of component.

From the definition of quality factor which is given above, the Q factor can be mathematically expressed as:

$$Q=\frac{E_{stored}}{E_{lost\,per\,cycle}}$$

When looking at the bandwidth of an RF resonant circuit this translates to:

$$Q=\frac{F_o}{F_{3dB}}$$

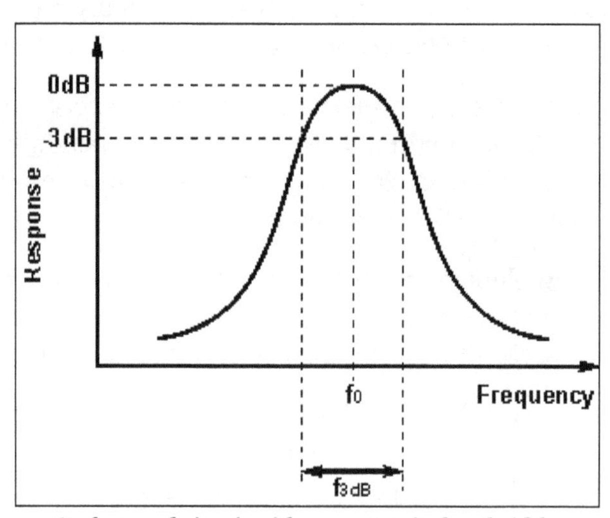

Q of a tuned circuit with respect to its bandwidth.

Within any RF or other circuit, each individual component can contribute to the quality factor of the circuit network as a whole. The Q of the components such as inductors and capacitors are often quoted as having a certain Q or quality factor.

Maxwell's Bridge

Using this bridge, we can measure the value of inductance by comparing it with a standard variable self-inductance arranged in bridge circuit as shown in the figure (a).

Consider Maxwell's inductance bridge as shown in the figure. Two branches consists of non-inductive resistances R_1 and R_2. One of the arms consists of variable inductance with series resistance r. The remaining arm consists of unknown inductance L_x.

At balance, we get condition as:

$$\frac{R_1}{\left[(R_3+r)+j\omega L_3\right]}=\frac{R_2}{R_X+j\omega L_X} \quad ...(1)$$

$$\Rightarrow R_1\left[R_X+j\omega L_X\right]=R_2\left[(R_3+r)+j\omega L_3\right]$$

$$\Rightarrow R_1R_X+j\omega R_1L_X=R_2\left[R_3+r\right]+j\omega R_2\,L_3$$

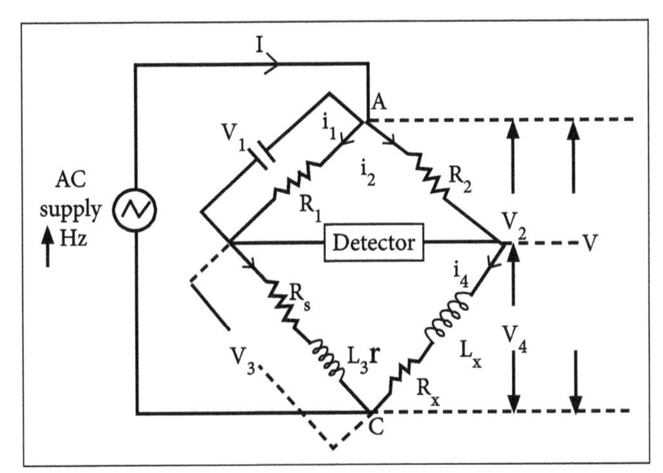

Circuit diagram.

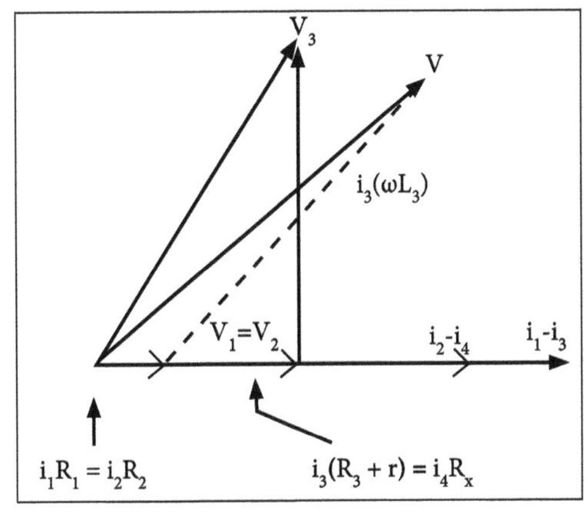

Phasor diagram.

Equating the imaginary terms, we can write as:

$$R_1 R_X = R_2 L_3$$

$$\therefore L_X = \frac{R_2}{R_1} L_3 \qquad ...(2)$$

Equating real terms, we can write:

$$R_1 R_X = R_2 \left(R_3 + r\right)$$

$$\therefore R_X = \frac{R_2}{R_1}\left(R_3 + r\right) \qquad ...(3)$$

Under the balanced condition, the vector diagram for Maxwell's inductance bridge is shown in the figure.

Hay's Bridge

The Hay's bridge is modification of the Maxwell's bridge. The connection of the Hay's bridge is shown in figure below. This Hay's bridge uses a resistor in series with the standard capacitor (unlike the Maxwell's bridge which uses a resistance in parallel with the capacitor).

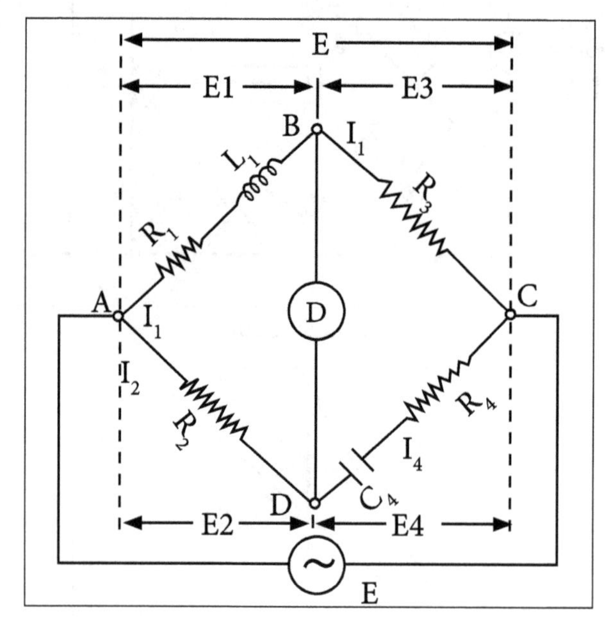

Hay's bridge for measurement of inductance.

Let,

L_1 = unknown resistance having a resistance R_1,

R_2, R_3, R_4 = known non-inductive resistance,

C_4 = standard capacitor.

At balance:

$$(R_1 + jwL_1)\left(R_4 - \frac{j}{wC_4}\right) = R_2R_3$$

$$R_1R_4 + \frac{L_1}{C_4} + jwL_1R_4 - \frac{jR_1}{wC_4} = R_2R_3$$

Separating real and imaginary term, we obtain:

$$R_1R_4 + \frac{L_1}{C_4} = R_2R_3 \text{ and } L_1 = \frac{R_1}{w^2R_4C_4}$$

Solving the above two equations we have:

$$L_1 = \frac{R_2R_3C_4}{1 + w^2C_4^2R_4^2}$$

And,

$$R_1 = \frac{w^2R_2R_3R_4C_4^2}{1 + w^2C_4^2R_4^2}$$

The Q factor of the coil is:

$$Q = \frac{wL_1}{R_1} = \frac{1}{wC_4R_4}$$

Advantages of the Hay's Bridge:

- This bridge gives very simple expression for the unknown inductance for high Q coils and is suitable for coils having $Q > 10$.
- This bridge also gives the simple expression for Q factor.
- From expression of Q factor it is clear that for high Q factor the value of resistance R4 should be small.

Disadvantages:

- The Hay's bridge is suited for measurement of high Q inductors, specially those inductors with $Q > 10$. For inductors having Q values smaller than 10, the term (1/Q)^2 in the expression for inductance L_1 becomes rather the important one and thus cannot be neglected. Hence this bridge is not suited for the measurement of coils having Q less than 10 and for these applications a Maxwell's bridge is more suited.

Anderson's Bridge

It is a modification of the Maxwell's inductance – Capacitance Bridge. Here, the self-inductance is measured in terms of a standard capacitor. The method is applicable for precise measurement of self-inductance over a very wide range of values.

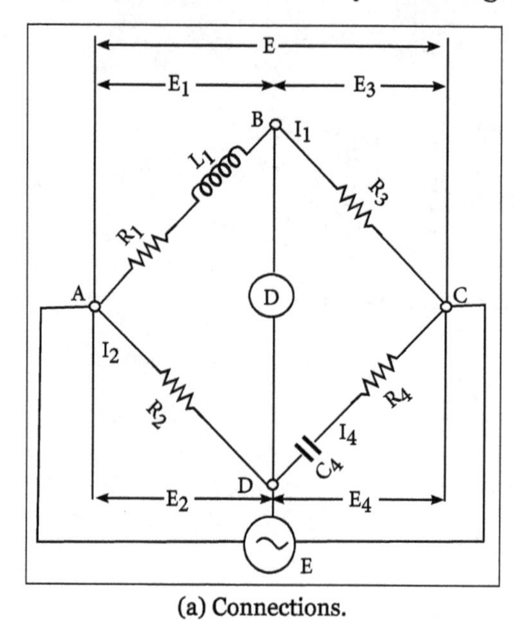

(a) Connections.

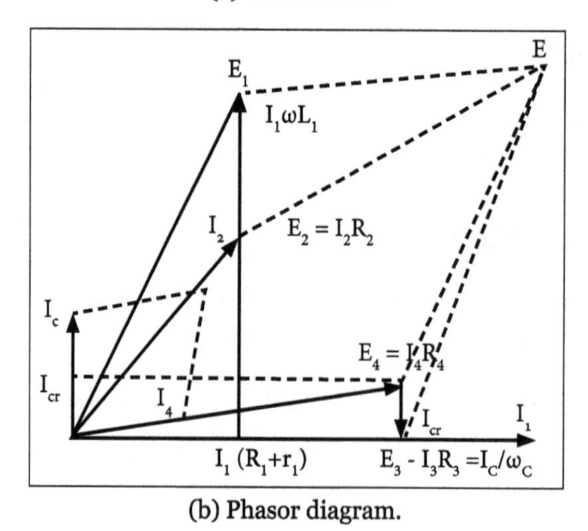

(b) Phasor diagram.

Let,

L_1 = Self-inductance to be measured.

R_1 = Resistance of self-inductor.

r_1 = Resistance connected in series with self-inductor.

r, R_2, R_3, R_a = Known non-inductive resistances.

C = fixed standard capacitor.

At balance, $I_1 = I_3$ and $I_2 = I_c + I_4$

$$\text{Now, } I_1 R_3 = I_C \times \frac{1}{j\omega C}$$

$$I_c = I_1 R_3 j\omega C.$$

Writing the other balance equations:

$$I_1(r_1 + R_1 + j\omega L_1) = I_2R_2 + I_cr$$

And,

$$I_C\left[r + \frac{1}{j\omega C}\right] = (I_2 - I_c)R_4.$$

Substituting the value of I_c in the above equations, we have:

$$I_1(r_1 + R_1 + j\omega L_1) = I_2R_2 + I_1 j\omega CR_3 r$$

Or,

$$I_1(r_1 + R_1 + j\omega L_1 - j\omega C R_3 r) = I_2R_2$$

And,

$$j\omega CR_3 I_1\left[r + \frac{1}{j\omega C}\right] = \left[I_2 - I_1 j\omega CR_3\right]R_4$$

Or,

$$I_1(j\omega CR_3 r + j\omega CR_3 R_4 + R_3) = I_2R_4 \quad ...(2)$$

From equations (1) and (2), we obtain:

$$I_1\left[r_1 + R_1 + j\omega L_1 - j\omega CR_3 r\right] = I_1\left[\frac{R_2 R_3}{R_4} + \frac{j\omega CR_2 R_3 r}{R_4} + j\omega CR_3 R_2\right].$$

Equating the real and the imaginary parts:

$$R_1 = \frac{R_2 R_3}{R_4} - r_1$$

And,

$$L_1 = C\frac{R_3}{R_4}\left[r(R_4 + R_2) + R_2 R_4\right] \qquad ...(4)$$

Advantages:

- It is much easier to obtain balance in the case of Anderson's bridge than in Maxwell's bridge for low Q-coils.
- A fixed capacitor can be used instead of a variable capacitor as in the case of max well's bridge.
- This bridge may be used for the accurate determination of capacitance in terms of inductance.

Disadvantages:

- It is more complex.
- Balance equations are more tedious.
- An additional junction point increases the difficulty of shielding the bridge.

1.4.1 Measurement of Mutual Inductance by Felici's Method, and as Self-inductance

Measurement of Mutual Inductance

In order to measure mutual inductance, the primary of the mutual inductor is connected to an a.c. source through an ammeter and the voltage induced in the secondary winding can be measured by using a voltmeter. However it should be ensured that supply voltage waveform is purely sinusoidal and no harmonics are there.

Let,

I = instantaneous value of current in the primary winding = $i_m \sin\omega t$

e = voltage induced in the secondary

Then the r.m.s value $E = M\omega I$ and $M = E/\omega I$ where $\omega = 2\pi f$ ...(3)

The accuracy of the method therefore, depends upon the accuracy with which the frequency of the source can be measured.

Felici's Method

It is the simplest a.c. bridge method for the measurement of mutual inductance. Only one standard, known and variable mutual inductor is needed for the measurement of the unknown. The diagram of connection is shown in Figure, where M_x, is the mutual inductor to be measured and M_1 is a variable standard mutual inductor. The two primary windings of the mutual inductors are connected in series across an a.c. source and the two secondary, windings are connected in series opposition. Secondary winding M_1 is varied until the vibration galvanometer shows zero deflection. At balance, the reading of M_s gives the value of the unknown. This simple method has several disadvantages such as, (i) the range of M_s should be more than the unknown. (ii) At high frequencies, the self-capacitance of the unknown, and the eddy currents induced in the metal parts may become so high that it is almost impossible to attain perfect balance of the bridge.

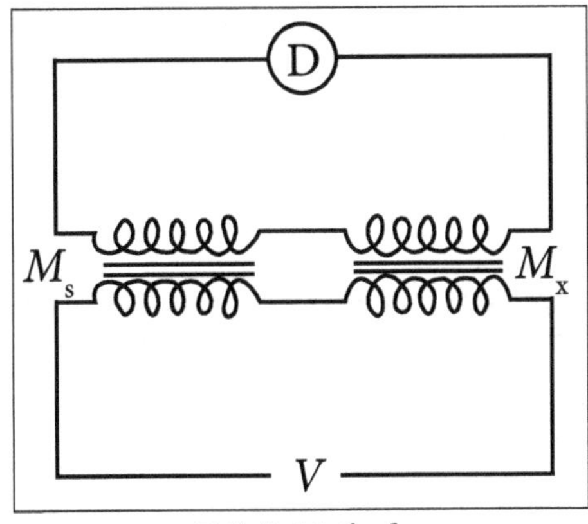

Felici's Method.

The Campbell Bridge

This is basically a modification of Felici's bridge and is designed for the comparison of two mutual inductances. Balance is obtained in the following two steps:

- First the bridge is balanced as a self-inductance bridge.
- Secondly it is balanced as a mutual-inductance bridge. The diagram of connections is shown in Figure.

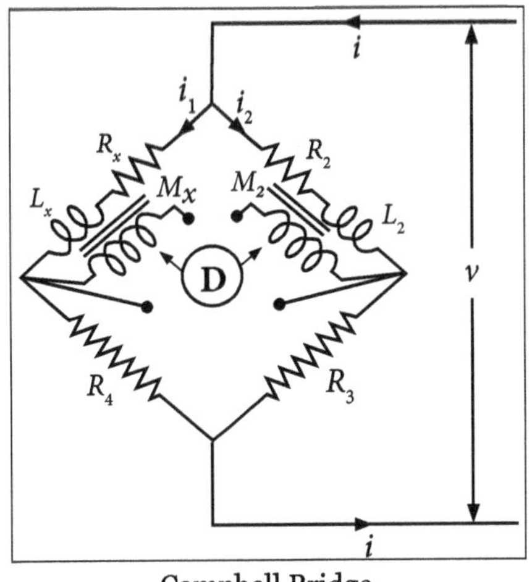

Campbell Bridge.

When balanced as a self-inductance bridge the balance conditions are:

$$i_1(R_x + j\omega L_x) = i_2(R_2 + j\omega L_2) \text{ and } i_1 R_4 = i_2 R_3$$

Then,

$$i_2/i_1 = L_x/L_2 = R_x/R_2 = R_4/R_3$$

$$\text{Then } R_3\left[R_x + j\omega(L_x - M_x)\right] = R_4\left[R_4 + j\omega(L_2 - M_2)\right]$$

$$\text{Therefore, } R_x/R_2 = R_4/R_3 = L_x/L_2 = M_x/M_2$$

From which the value of M_x can be determined.

1.5 Capacitance: Measurement of Capacitance

Capacitors are among the most useful of all electronic components and capacitance is a term that refers to the ability of a capacitor to store charge. It's also the measurement used to indicate how much energy a particular capacitor can store. The more capacitance a capacitor has, the more charge it can store.

Capacitance is measured in units called farads (abbreviated F). The definition of one farad is deceptively simple. A one-farad capacitor holds a voltage across the plates of exactly one volt when it's charged with exactly one ampere per second of current.

The "one ampere per second of current" part is really referring to the amount of charge present in the capacitor. There's no rule that says the current has to flow for a full second. It could be one ampere for one second, or two amperes for half a second, or half an ampere for two seconds or it could be 100 mA for 10 seconds or 10 mA for 100 seconds.

One ampere per second corresponds to the standard unit for measuring electric charge, called the coulomb. So another way of stating the value of one farad is to say that it's the amount of capacitance that can store one coulomb with a voltage of one volt across the plates.

It turns out that one farad is a huge amount of capacitance, simply because one coulomb is a very large amount of charge. To put it into perspective, the total charge contained in an average lightning bolt is about five coulombs and we need only five, one-farad capacitors to store the charge contained in a lightning strike. (Some lightning strikes are much more powerful, as much as 350 coulombs.)

It's a given that Doc Brown's flux capacitor was in the farad range because Doc charged it with a lightning strike. But capacitors used in electronics are charged from much more modest sources.

In fact, the largest capacitors we're likely to use have capacitance that is measured in millionths of a farad, called microfarads and abbreviated μF and the smaller ones are measured in millionths of a microfarad, also called a Pico farad and abbreviated pF.

Capacitor Measurements

Like resistors, capacitors aren't manufactured to perfection. Instead, most capacitors have a margin of error, called tolerance. In some cases, the margin of error may be as much as 80%. Fortunately, that degree of impression rarely has noticeable effect on most circuits.

The μ in μF isn't an italic letter u; it's the Greek letter mu, which is the common abbreviation for micro.

It's common to represent values of 1,000 pF or more in μF rather than pF. For example, 1,000 pF is written as 0.001 μF and 22,000 pF is written as 0.022 μF.

Loss Angle

A pure insulator when connected across line and earth, behaves as a capacitor. In an ideal insulator, as the insulating material that acts as dielectric too, is 100 % pure, the electric current passing through the insulator, only has capacitive component. There is no resistive component of the current, flowing from line to earth through insulator as in case of ideal insulating material, there is zero percent impurity. In pure capacitor, the capacitive electric current leads applied voltage by 90°. In practice, the insulator cannot be made 100% pure. Also due to the ageing of insulator the impurities like dirt and moisture enter into it. These impurities provide the conductive path to the current. Consequently, leakage electric current flowing from the line earth through insulator has also resistive component.

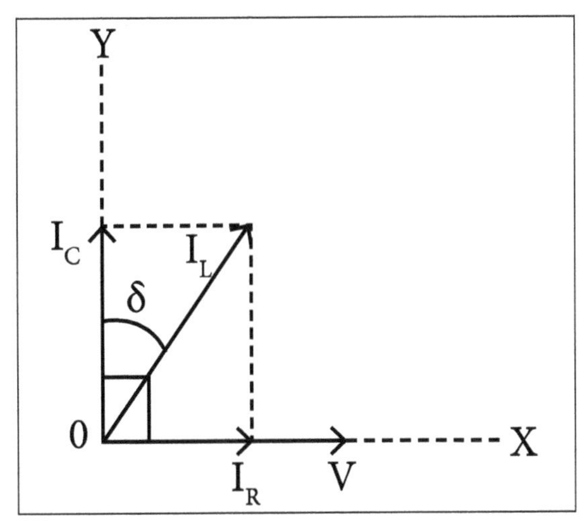

Hence, it is needless to say that, for a good insulator, this resistive component of leakage electric current is quite low. In other way the healthiness of an electrical insulator can be determined by ratio of resistive component to capacitive component. For a good insulator this ratio would be quite low. This ratio is commonly known as tan δ or tan delta. Sometimes it is also referred as dissipation factor.

In the vector diagram above, the system voltage is drawn along x-axis. Conductive electric current i.e. the resistive component of the leakage current, I_R will also be along x-axis. As the capacitive component of the leakage electric current I_C leads system voltage by 90°, it will be drawn along y-axis. Now, total leakage electric current $I_L(I_C + I_R)$ makes an angle δ (say) with the y-axis. Now, from the diagram above, it is clear the ratio, I_R to I_C is nothing but tan δ or tan delta.

$$\text{Thus,} \quad \tan\delta = \frac{I_R}{I_C}$$

This δ angle is known as loss angle.

De Sauty's Bridge

This bridge provide us the most suitable method for comparing the two values of capacitor if we neglect dielectric losses in the bridge circuit. The circuit of De Sauty's bridge is shown below,

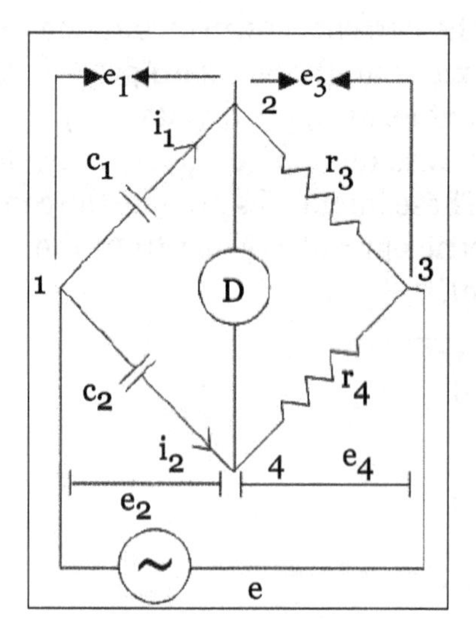

Battery is applied between terminals marked as 1 and 4. The arm 1-2 consists of capacitor c1 (whose value is unknown) which carries current i_1 as shown, arm 2 - 4 consists of pure resistor (here pure resistor means we are assuming it non inductive in nature), arm 3 - 4 also consists of pure resistor and the arm 4 - 1 consists of the standard capacitor whose value is already known to us. Let us derive the expression for capacitor c_1 in terms of standard capacitor and resistors. At balance condition we have:

In order to obtain the balance point we must adjust the values of either r_3 or r_4 without disturbing any other element of the bridge. This is the most efficient method of comparing two values of capacitor if all the dielectric losses are neglected from the circuit.

Now let us draw and study phasor diagram of this bridge. Phasor diagram of De Sauty bridge is shown below:

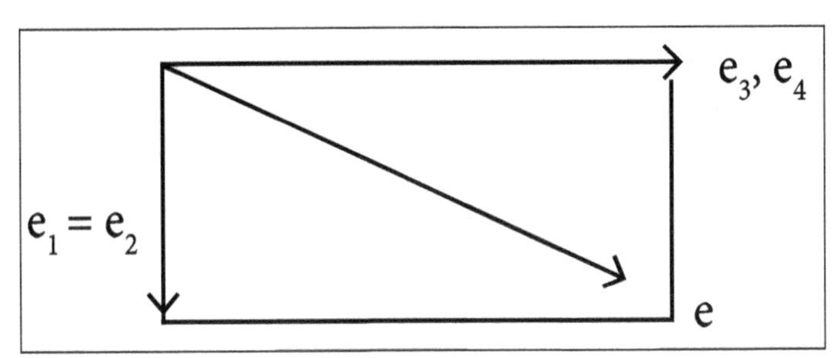

Let us mark the voltage drop across unknown capacitor as e_1, voltage drop across the resistor r_3 be e_3, voltage drop across arm 3 - 4 be e_4 and voltage drop across arm 4 - 1 be e_2. At balance condition the current flows through 2 - 4 path will be zero and also voltage drops e_1 and e_3 be equal to voltage drops e_2 and e_4 respectively. In order to draw the phasor diagram we have taken e_3 (or e_4) reference axis, e_1 and e_2 are shown at right angle to e_1 (or e_2).

Now instead of some advantages like bridge is quite simple and it provides easy calculations, there are some disadvantages of this bridge because this bridge give inaccurate results for imperfect capacitor (here imperfect means capacitors which not free from dielectric losses). Hence we can use this bridge only for comparing perfect capacitors. Here we modify the De Sauty's bridge; we want to have such a kind of bridge which will gives us accurate results for imperfect capacitors also. This modification is done by Grover. The modified new circuit diagram is shown below:

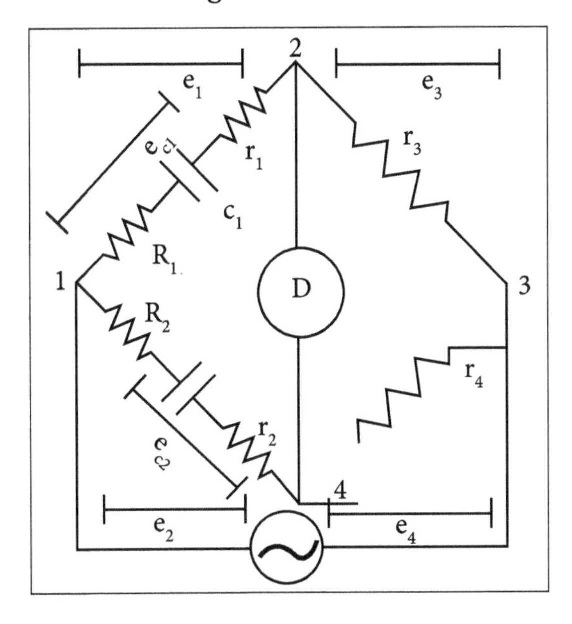

Here Grover has introduced electrical resistances r_1 and r_2 as shown above on arms 1 - 2 and 4 - 1 respectively, in order to include the dielectric losses. Also he has connected resistances R_1 and R_2 respectively in the arms 1 - 2 and 4 - 1. Let us derive the expression capacitor c_1 whose value is unknown to us. Again we connected standard capacitor on the same arm 1 - 4 as we have done in De Sauty's bridge.

By drawing the phasor diagram we can calculate dissipation factor. Phasor diagram for the above circuit is shown below:

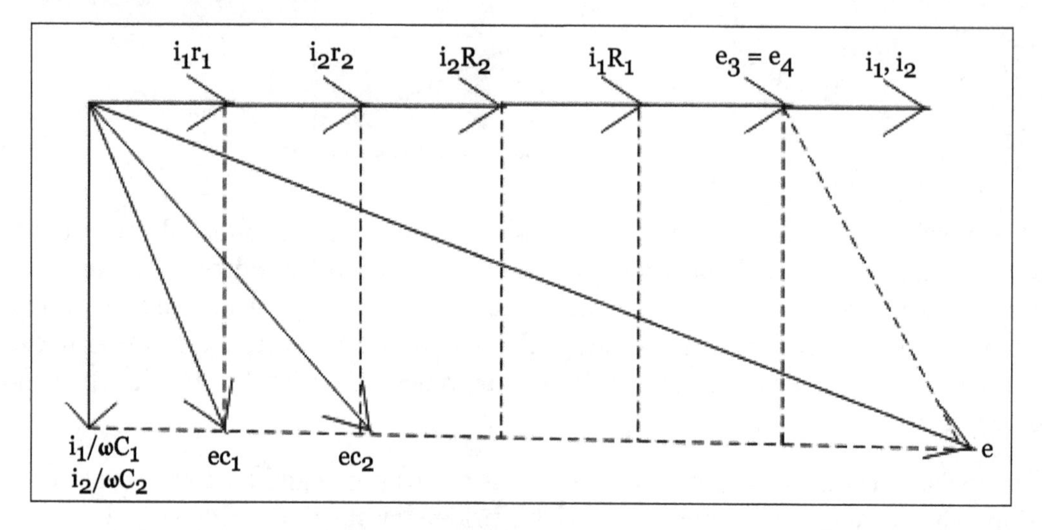

Let us mark δ_1 and δ_2 be phase angles of the capacitors c_1 and c_2 capacitors respectively. From the phasor diagram we have $\tan(\delta_1)$ = dissipation factor $= \omega c_1 r_1$ and similarly we have $\tan(\delta_2) = \omega c_2 r_2$. From the equation we have:

$$c_2 r_2 - c_1 r_1 = c_1 R_1 - c_2 R_2$$

On multiplying ω both sides we have:

$$\omega c_2 r_2 - \omega c_1 r_1 = \omega(c_1 R_1 - c_2 R_2)$$

$$\text{But } \frac{c_1}{c_2} = \frac{r_4}{r_3}$$

Therefore the final expression for the dissipation factor is written as:

$$\tan(\delta_1) - \tan(\delta_2) = \omega c_2\left(R1\frac{r_4}{r_3} - R_2\right)$$

Hence if dissipation factor for one capacitor is known, same for the other can be found. However this method gives quite inaccurate results for dissipation factor.

Schering Bridge

This bridge is used to measure the capacitance of the capacitor, dissipation factor and measurement of relative permittivity.

Let us consider the circuit of Schering bridge as shown below:

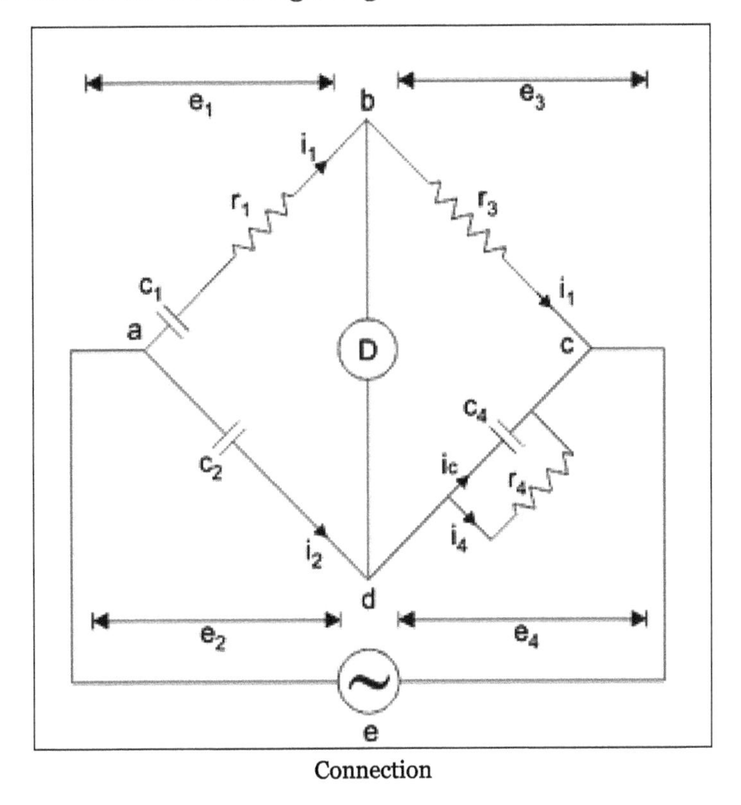

Connection

Here c_1 is the unknown capacitance whose value is to be determined with the series electrical resistance r_1.

c_2 is a standard capacitor.

c_4 is a variable capacitor.

r_3 is the pure resistor (i.e. non inductive in nature) and r_4 is a variable non inductive resistor connected in parallel with variable capacitor c_4.

Now the supply is given to the bridge between the points a and c. The detector is connected between b and d.

Using the balancing equation we have:

$$\left(r_1 + \frac{1}{j\omega c_1}\right) r_4 = \frac{r_3}{j\omega c_2}\left(1 + j\omega c_4 r_4\right)$$

$$r_1 r_4 - \frac{jr_4}{\omega c_1} = -\frac{jr_3}{\omega c_2} + \frac{r_3 r_4 c_4}{c_2}$$

Equating the real and imaginary parts and then separating we get:

$$r_1 = \frac{r_3 c_4}{c_2}$$

$$c_1 = c_2 \frac{r_4}{r_3}$$

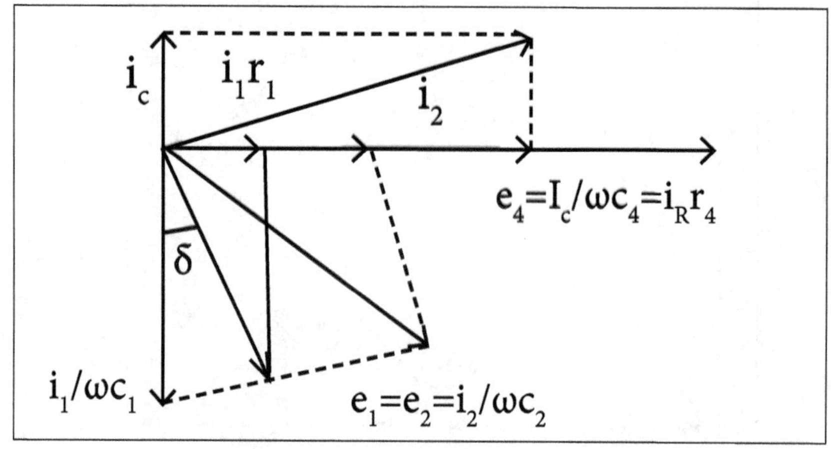

Phasor diagram.

Let us consider the phasor diagram of the above Schering bridge circuit and mark the voltage drops across ab, bc, cd and ad as e_1, e_3, e_4 and e_2 respectively.

From the above Schering bridge phasor diagram we can calculate the value of tan δ which is also called the dissipation factor.

$$\tan\delta = \omega c_1 r_1 = \omega \frac{c_2 r_4}{r_3} \times \frac{r_3 c_4}{c_2} = \omega c_4 r_4$$

The equation that we have derived above is quite simple and dissipation factor can be calculated easily.

Simple Schering bridge (which uses low voltages) is used for measuring the dissipation factor, capacitance and measurement of other properties of insulating materials like insulating oil etc.

Need of High Voltage Schering Bridge

For the measurement of small capacitance we need to apply high voltage and high frequency as compared to low voltage which suffers many disadvantages.

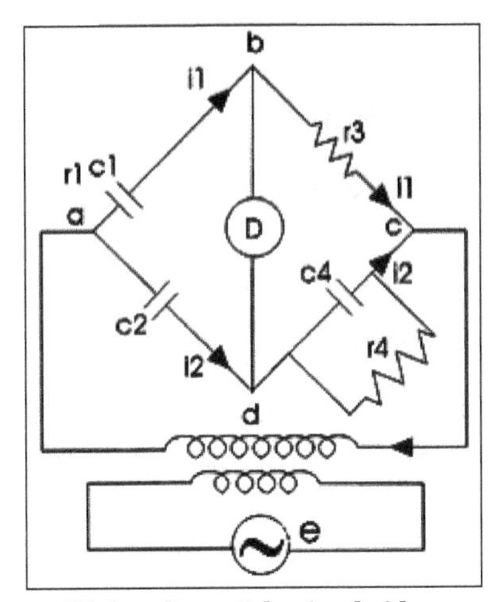

High voltage Schering bridge.

Features of this High Voltage Schering Bridge

- The two bridge arms ab and ad has only capacitors as shown, the bridge given below and impedances of these two arms are quite large as compared to the impedances of bc and cd.

The arms bc and cd contains resistor r_3 and parallel combination of capacitor c_4 and resistor r_4 respectively. As impedances of bc and cd are quite small therefore drop across bc and cd is small. The point c is earthed, so that the voltage across bc and dc are of few volts above the point c.

- The high voltage supply is obtained from a transformer of 50 Hz and the detector in this bridge is a vibration galvanometer.
- The impedances of arms ab and ad are large therefore this circuit draws low current hence power loss is low but due to this low current we need a very sensitive detector to detect this low current.
- The fixed standard capacitor c_2 has compressed gas which works as dielectric therefore dissipation factor can be taken as zero for compressed air.

Earthed screens are placed between the high and the low arms of the bridge which prevents the errors caused due to inter-capacitance.

Let us study how Schering bridge measures relative permittivity: In order to measure the relative permittivity first we need to measure capacitance of a small capacitor with specimen as dielectric and from this the measured value of capacitance relative permittivity can be calculated easily by using the very simple relation.

$$r = \frac{cd}{\varepsilon A}$$

Where, r is relative permeability.

c is the capacitance with specimen as dielectric.

d is the spacing between the electrodes.

A is the net area of electrodes.

And ε is permittivity of free space.

There is another way to calculate relative permittivity of the specimen by changing electrode spacing. Let us consider the diagram shown below:

Here,

A is the area of electrode.

d is the thickness of the specimen.

t is the gap between the electrode and specimen.

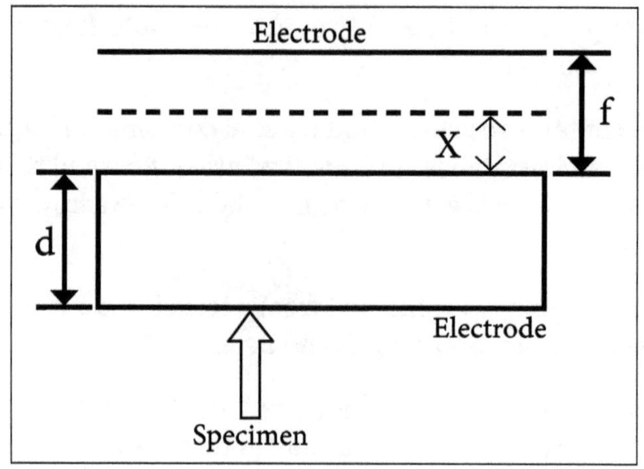

Calculating the relative permittivity of specimen.

c_s is the capacitance of specimen.

c_o is capacitance due to spacing between electrode and specimen.

c is the effective combination of c_s and c_o.

$$\text{Now,}\quad c = \frac{c_s c_o}{c_s + c_o}$$

From figure above two capacitors are connected in series.

$$c = \frac{\epsilon_r \epsilon_0 A}{\epsilon_r t + d}$$

εo is permittivity of free space, εr is relative permittivity.

When we remove specimen and the spacing is readjusted to have the same value of capacitance, the expression for capacitance reduces to:

$$c = \frac{\epsilon_0 A}{t+d-x} \qquad ....(2)$$

On equating the equations (1) and (2), we will get the final expression for ε_r as:

$$\varepsilon_r = \frac{d}{d-x}$$

The necessary balance conditions for a Schering bridge is:

$$r_1 R_4 - \frac{jR_4}{\omega C_1} = -\frac{jR_3}{\omega C_2} + \frac{R_3 R_4 C_4}{C_2}$$

Equating the real and imaginary terms:

$$\gamma_1 = \frac{R_3 R_4}{C_2}$$

$$C_1 = C_2 = \left(\frac{R_4}{R_3}\right)$$

Wien's Bridge

Basically the bridge is used for the frequency measurement but it is also used for the measurement of unknown capacitor with greater accuracy.

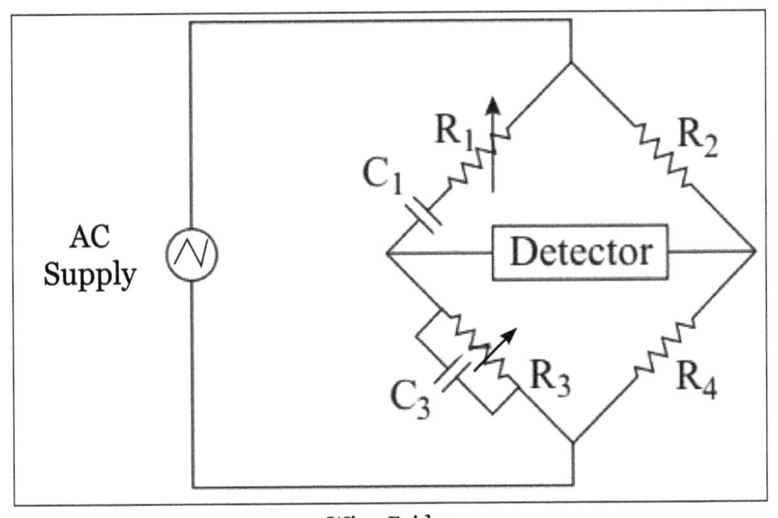

Wien Bridge

It's one ratio arm consists of a series RC circuit, i.e., R_1 and C_1. The second ratio arm consists of a resistance R_2. The third arm consists of the parallel combination of resistance and capacitor i.e., R_3 and C_3.

From the figure, we can write:

$$Z_1 = R_1 - j\left(\frac{1}{\omega C_1}\right)$$

$$Z_2 = R_2$$

$$Z_3 = R_3 \| C_3 ; Y_3 = \frac{1}{R_3} + j\omega C_3.$$

$$Z_4 = R_4.$$

The balance condition is:

$$\overline{Z_1 Z_4} \quad \overline{Z_2 Z_3}$$

$$\overline{Z_2} = \frac{\overline{Z_1 Z_4}}{\overline{Z}_3} = Z_1 \ \overline{Z_4 Y_3}$$

$$\therefore R_2 = \left[R_1 - j\left(\frac{1}{\omega_{c1}}\right)\right] R_4 \left[\frac{1}{R_3} + j\omega_{c3}\right]$$

$$= R_4 \left[\frac{R_1}{R_3} + j_{\omega} R_1 C_3 - j\frac{1}{\omega_{c1} R_3} + \frac{C_3}{C_1}\right]$$

$$= R_4 \left[\frac{R_1}{R_3} + \frac{C_3}{C_1}\right] + jR_4 \left[\omega R_1 C_3 - \frac{1}{\omega C_1 C_3}\right]$$

Equating real parts of both sides:

$$R_2 = \frac{R_4 R_1}{R_3} + \frac{C_3 R_4}{C_1}$$

$$\frac{R_2}{R_4} = \frac{R_1}{R_3} + \frac{C_3}{C_1}.$$

Equating imaginary parts of both sides:

$$\omega R_1 C_3 - \frac{1}{\omega C_1 R_3} = 0$$

$$\omega^2 = \frac{1}{R_1 R_3 C_1 C_3}$$

$$\omega = \frac{1}{\sqrt{R_1 R_3 C_1 C_3}}$$

$$f = \frac{1}{2\pi\sqrt{R_1 R_3 C_1 C_3}}.$$

$$\text{Generally, } R_1 = R_3 = R \text{ and } C_1 = C_3 = C.$$

$$\therefore F = \frac{1}{2\pi RC}$$

1.5.1 Screening of Bridge Components and Wagnor Earthing Device

Screening of AC Bridges

Errors may be introduced in bridge measurements due to imperfect insulation and residuals in the components, due to magnetic couplings between the components, due to electrostatic couplings between the components, and between the components and the earth. Under high frequency and high voltage conditions, the errors become significant. The errors due to first three causes can be reduced by using high quality components, co-axial connecting leads, and laying out the bridge in such a manner that the least possible loop areas are formed.

The electrostatic couplings introduce several stray capacitances. To reduce the errors caused by these capacitances, we use screened components, screen the whole bridge, and provide Wagner Earth. Screening renders the capacitances definite in value. Further, screening also puts several of the stray capacitances effectively across the source or across the detector when no error can be caused. Screening can place some of the capacitances across the standard components where these can be accounted for in calibration.

When the capacitances are so reduced that these appear between the corner points A, B, C, and D of the bridge and the earth, and if one terminal of the detector branch is earthed; the capacitances C_A and C_c only can cause error. The errors due to this also can be eliminated by using Wagner earth, wherein the auxiliary arms Z_5 and Z_6, of the same type as Z_2 and Z_4, are provided. Instead of ear thing point B, the point E, the junction point of Z_5 and Z_6 is earthed.

The bridge is balanced with Z_2-Z_4, and Z_5 - 2_6 in circuit alternately, i.e., the detector connected to B, or E alternately. This way, the points B and D are brought to earth potential without actually connecting either of these to earth. The stray earth capacitances C_A and C_c are now across Z_5, and Z_6; and not across Z_2, and Z_4.

The errors can also be eliminated from the measurement by applying the difference method, or the substitution method. In the difference method, two measurements are taken-one with the unknown in position, and the other with unknown removed and the terminals left open if the unknown is a capacitor and shorted if the unknown is an inductor or a resistor. The difference of the two measurements taken is the error-free value of the unknown.

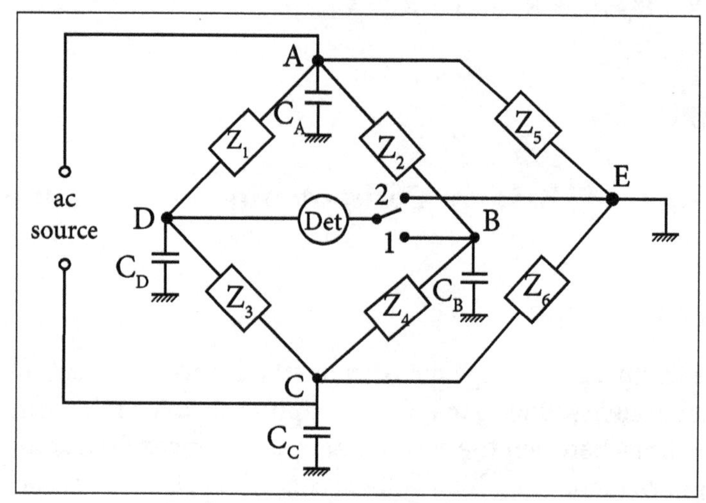

Earth capacitances and wagner earth.

In the substitution method the second measurement is taken by substituting the unknown by a variable standard of the same type and the bridge is balanced by varying the value of this standard component. The balance value of the standard is the value of the unknown.

Wagner's Earthing Device

Directly connecting the null detector to ground is not an option, as it would create a direct current path for stray currents, which would be worse than any capacitive path. Instead, a special voltage divider circuit called a Wagner ground or Wagner earth may be used to maintain the null detector at ground potential without the need for a direct connection to the null detector.

The Wagner earth circuit is nothing more than a voltage divider, designed to have the voltage ratio and phase shift as each side of the bridge. Because the midpoint of the Wagner divider is directly grounded, any other divider circuit including either side of the bridge having the same voltage proportions and phases as the Wagner divider and powered by the same AC voltage source, will be at ground potential as well. Thus, the

Wagner earth divider forces the null detector to be at ground potential, without a direct connection between the detector and ground.

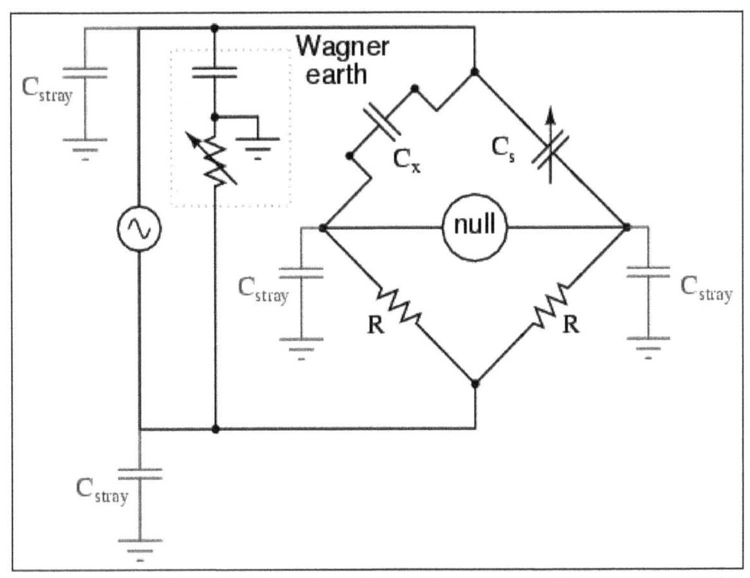

Wagner ground for AC supply minimizes the effects of stray capacitance to ground on the bridge.

There is often a provision made in the null detector connection to confirm proper setting of the Wagner earth divider circuit: a two-position switch, (Figure below) so that one end of the null detector may be connected to either the bridge or the Wagner earth. When the null detector registers zero signals in both switch positions, the bridge is not only guaranteed to be balanced, but the null detector is also guaranteed to be at zero potential with respect to ground, thus eliminating any errors due to leakage currents through stray detector-to-ground capacitances:

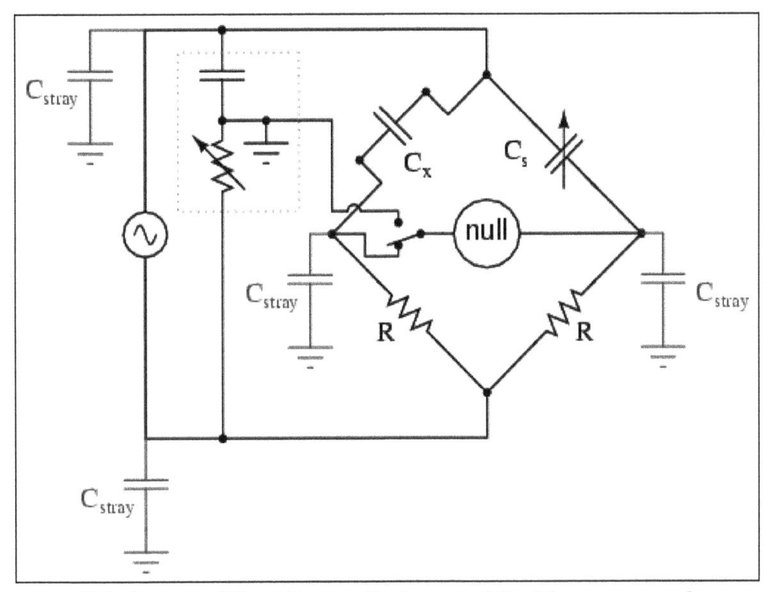

Switch-up position allows adjustment of the Wagner ground.

2

Types of Electrical Meters and Power Factor

2.1 Galvanometer

Ballistic Galvanometer

The ballistic galvanometer is a suspended moving coil galvanometer with very small damping. It is used for measuring charge by observing the maximum deflection in its oscillatory motion. For this purpose the time for the charge to pass through the galvanometer should be short compared to the period of oscillation of the coil.

The total effective damping b of the galvanometer is given by:

$$b = b_m + C / R^2$$

Where b_m is the damping due to the mechanical part, C a constant, and R the total resistance in the circuit. The second term in the right hand side of equation above arises out of induced emf produced in the coil as it moves in the magnetic field of the permanent magnets. This impedes the motion of the coil, and is called electromagnetic damping. This can be reduced by increasing R. When the coil oscillates with the open circuit, R is infinity and electromagnetic damping vanishes.

When a charge Q is passed through the galvanometer, its effect is an impulse on the coil. The rotation of the coil through an angle q which is proportional to the charge Q. A mirror fixed on the plane of the coil reflects light from a lamp and scale arrangement is kept 1m away from the galvanometer on the scale of the arrangement. The deflection d1 of the light spot on the scale is proportional to rotation of the coil for small angle of rotation. The relationship between the charge Q and d_1 is given by:

$$Q = \left[KT / (2p)\right] d_1 \ [\text{coulomb}]$$

Where T is the time period of oscillation of the coil and K, the figure of merit of the galvanometer:

$$K = k / (NAB2L)$$

Where k is torque constant of the wire of suspension of the coil, N the number of turns, A is the area, B the magnetic field produced by the magnets, and L the distance between lamp and the galvanometer (1 m). The figure of merit of the galvanometer is obtained by taking the ratio of steady current i and the corresponding deflection of the light spot.

The actual deflection produced is smaller due to the presence of damping. The size of the correction can be found by measuring the logarithmic decrement, l of the successive swings on the same side.

$$l = \ln(q_1/q_3) = \ln(q_3/q_5) = \ldots$$

$$l = \ln(d_1/d_3) = \ln(d_3/d_5) = \ldots$$

Corrected deflection d_o is given by:

$$d_o = d_1 e^{l/4}$$

Or since l is small:

$$d_o = d_1(1 + l/4) = d_1 + (d_1 - d_3)/4$$

The ability of ballistic galvanometer to measure the charge accurately, is employed for measuring high resistances, capacitances, magnetic field etc.

D' Arsonval Galvanometer

This instrument is very commonly used in various methods of resistance measurement and also in d. c. potentiometer work.

Construction of D' Arsonval galvanometer:

The construction of D' Arsonval galvanometer is shown in figure below,

Moving Coil

It is the current carrying element. It is rectangular or circular in shape and consists of number of turns of fine wire. This coil is suspended so that it is free to turn about its vertical axis of the symmetry. It is arranged in a uniform, radial, horizontal magnetic field in the air gap between the pole pieces of a permanent magnet and iron core. The iron core is spherical in shape if the coil is circular but is cylindrical if the coil is rectangular. The iron core is used to provide a flux path of low reluctance and therefore to provide the strong magnetic field for the coil to move in, this increases the deflecting torque and hence sensitivity of the galvanometer. The length of air gap is about 1.5 mm. In some galvanometers the iron core is omitted resulting in decreased value of flux density and the coil is made narrower to decrease air gap. Such galvanometer is less sensitive, but its moment of inertia is smaller on account of its reduced radius and consequently a short periodic time.

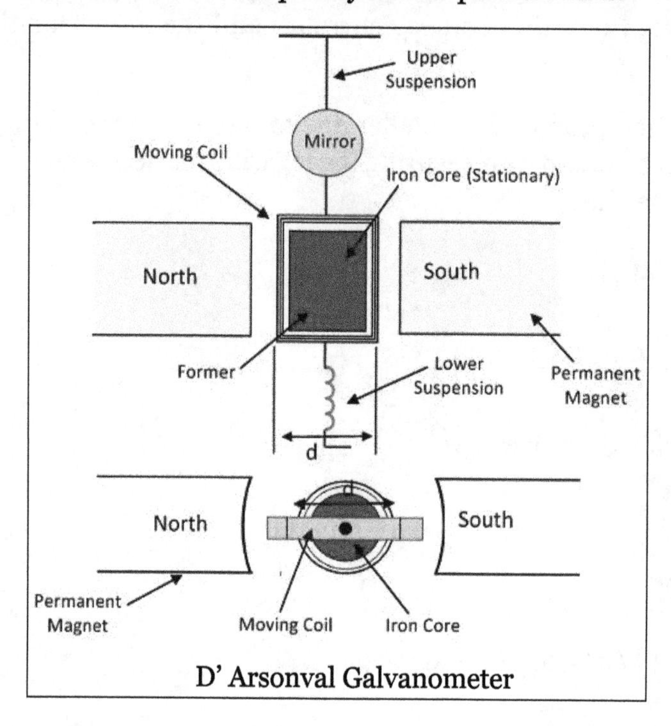

D' Arsonval Galvanometer

Damping

There is a damping torque present owing to production of eddy currents in the metal former on which coil is mounted. Damping is also obtained by connecting a low resistance across galvanometer terminals. Damping torque depends upon the resistance and we can obtain critical damping by adjusting the value of resistance.

Suspension

The coil is supported by a flat ribbon suspension which also carries current to coil. The other current connection in a sensitive galvanometer is a coiled wire. This is called the lower suspension and it has a negligible torque effect. This type of galvanometer must be leveled carefully so that the coil hangs straight and centrally without rubbing poles or the soft iron cylinder. Some portable galvanometers which does not require exact leveling have" taut suspensions" consisting of the straight flat strips kept under tension at the both top and at the bottom.

The upper suspension consists of gold or copper wire of nearly 0.012-5 or 0.02-5 mm diameter rolled in the form of a ribbon. This is not very strong mechanically; so that the galvanometers must he handled carefully without jerks. Sensitive galvanometers are provided with the coil clamps to the strain from suspension, while the galvanometer is being moved.

Indication

The suspension carries a small mirror upon that a beam of light is cast. The beam of light which is reflected on a scale upon which deflection is measured. This scale is usually about 1 meter away from instrument, although ½ meter may be used for greater compactness.

Zero Setting

A torsion head is provided for adjusting the position of coil and also for zero setting.

Vibration Galvanometer

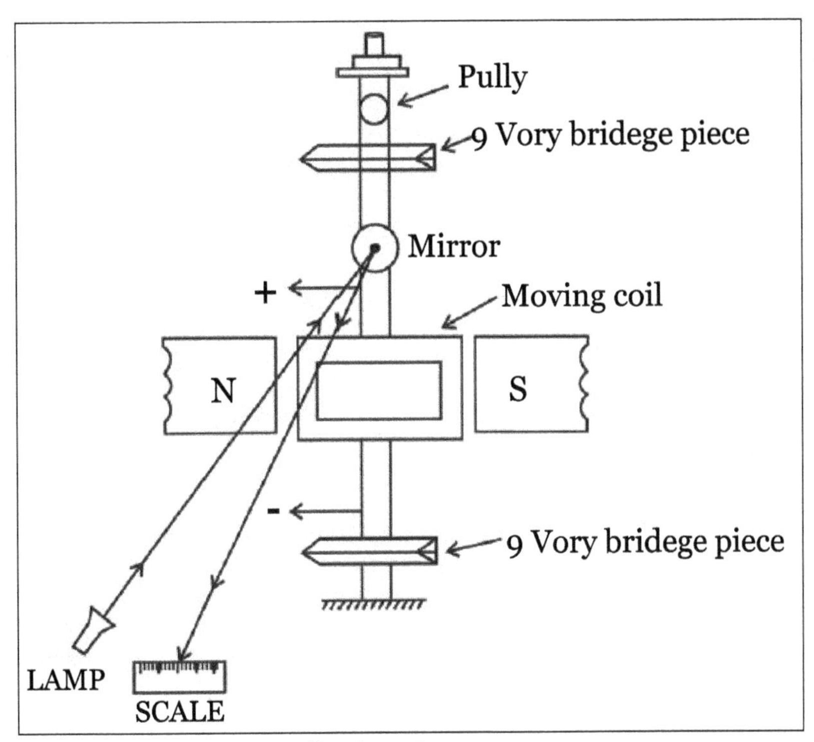

Vibration galvanometer

The construction of this galvanometer is similar to the PMMC instrument except for the moving system. The moving coil is suspended using two ivory bridge pieces. The tension of the system can be varied by rotating the screw provided at the top suspension. The natural frequency can be varied by varying the tension wire of the screw or varying the distance between ivory bridge piece. When A.C. current is passed through coil an alternating torque or vibration is produced. This vibration is maximum if the natural frequency of moving system coincide with supply frequency. Vibration is maximum, when resonance takes place. When the coil is vibrating the mirror oscillates and the dot moves back and front. This appears as a line on the glass scale. Vibration galvanometer is used for null deflection a dot appears on the scale. If the bridge is unbalanced, a line appears on the scale.

2.1.1 Influence of Resistance on Damping, Logarithmic Decrement, Calibration of Galvanometers, Galvanometer Constants, Measurement of Flux and Magnetic Field by using Galvanometers

Galvanometer Constants

The performance of galvanometer can be described in terms of two sets of constants namely the design constants (J, moment of inertia; D, damping constant; C, restoration constant; and G, deflection constant) or the working constants (S, sensitivity; R_g, galvanometer resistance; R_c, critical damping resistance or γ, relative damping and TO, free period). In case the design constants are known, the behavior of the galvanometer can be described completely for any condition of use but these constants are not usually known accurately and in most of the cases, it is very tedious to determine them. Hence the performance of galvanometer is usually expressed in terms of working constants. Various working constants are defined as below:

Current Sensitivity: The deflection caused by the unit current is known as current sensitivity. In case of a galvanometer attached with a mirror it may be defined as deflection produced in mm on a scale at a distance of one meter from the galvanometer mirror by unit current.

i.e. current sensitivity d/(i ×106) mm per micro-ampere where d is the deflection on scale in mm.

Sometimes the current sensitivity is defined as current required to cause a deflection of one scale division.

Voltage Sensitivity: The voltage sensitivity is defined in terms of the deflection in scale division caused by unit voltage applied across galvanometer

In case of a galvanometer attached with a mirror the voltage sensitivity is defined as the deflection in mm produced on a scale at a distance of 1 m from mirror by unit voltage impressed on the galvanometer i.e. Voltage sensitivity = d/(V × 106) mm/μv.

Magohm Sensitivity: The resistance required in MΩ placed in series with the galvanometer through which one volt of the impressed voltage will cause a deflection of 1mm on a scale at a distance of 1m from the mirror. This is reciprocal of current sensitivity.

Ballistic Sensitivity: It is defined as the first maximum deflection in mm caused on a scale placed at distance of 1 m from the mirror by a unit charge suddenly passed in a galvanometer.

Effect of External Resistance on Damping

The damping means opposition to the motion by dissipating the energy of rotation. In galvanometer the damping is provided by two types,

- Mechanical damping: This is due to the friction present in the mechanical motion of pointer. This is not very significant: The damping torque produced due to such mechanical effects is given by,

$$T_m = D_m \frac{D\theta}{dt}$$

Where,

$$D_m = \text{Mechanical damping constant} \qquad ...(1)$$

- Electromagnetic damping: This is effective damping than the mechanical damping. It is produced due to the induced effects when coil moves in a magnetic field. Thus when coil moves in a magnetic field,
    - The eddy currents are induced in the metal former.
    - The e.m.f. is induced in coil which circulates current through coil.

These two effects cause damping called electromagnetic damping.

Let R_g = Resistance of galvanometer circuit = $R_g + R_x$.

Where, R_g = Resistance of galvanometer coil.

R_x = External resistance connected for damping.

When coil rotates, emf is induced in it which is given by:

$$e = 2N \times Blv \qquad ...(2)$$

Where,

$$v = \frac{r}{2}\omega = \frac{r}{2}\frac{d\theta}{dt} = \text{linear velocity}$$

$$\therefore \quad e = 2NBl\frac{r}{2}\frac{d\theta}{dt} = NBA\frac{d\theta}{dt} \qquad ...l \times r$$

But $G = NBA$

$$\therefore \quad e = G\frac{d\theta}{dt} \qquad ...(3)$$

$$\therefore \quad i = \frac{e}{R} = \frac{G}{R}\frac{d\theta}{dt} \qquad ...(4)$$

The torque produced due to this current flowing through the coil is:

$$T_{coil} = N \times Bil \times r = NBAi = Gi$$

$$T_{coil} = G \times \frac{G}{R}\frac{d\theta}{dt} = \frac{G^2}{R}\frac{d\theta}{dt} = D_{coil}\frac{d\theta}{dt} \qquad ...(5)$$

where,

D_{coil} = Damping constant of coil circuit

$$\therefore \quad D_{coil} = \frac{G^2}{R} \qquad ...(6)$$

Now let us find damping due to the metal former. It consists of one strip i.e. N=1.

$$T_f = Bil \times r = BAi \qquad ....(7)$$

Now,

$$BA = \frac{G}{N} \text{ and } i = \frac{G}{NR_f}\frac{d\theta}{dt} \qquad \text{...as } N = 1 \text{ for former}$$

where,

R_f = Resistance of former

Therefore,

$$T_f = \frac{G}{N} \times \frac{G}{NR_f}\frac{d\theta}{dt} = \frac{G^2}{N^2R_f}\frac{d\theta}{dt} = D_{former}\frac{d\theta}{dt} \qquad ...(8)$$

where,

D_{former} = Damping constant of former

Therefore,

$$D_{former} = \frac{G^2}{N^2R_f} \qquad ...(9)$$

Total electromagnetic damping $= \left(D_{coil} + D_{former}\right)\frac{d\theta}{dt}$

Therefore,

$$T_e = \left[\frac{G^2}{R} + \frac{G^2}{N^2R_f}\right]\frac{d\theta}{dt}$$

Therefore,

$$T_e = D_e \frac{d\theta}{dt} \qquad ...(10)$$

Where,

$$D_e = \left[\frac{G^2}{R} + \frac{G^2}{N^2 R_f}\right] = \text{damping constant due to electromagnetic damping} \qquad ...(11)$$

The total damping due to both effects is:

$$T_D = T_m + T_e = [D_m + D_e]\frac{d\theta}{dt} = D\frac{d\theta}{dt} \qquad ...(12)$$

where $D = D_m + D_2$

Logarithmic Decrement

There are many methods for measuring the damping of a vibration system. Logarithmic decrement method and bandwidth method are introduced here.

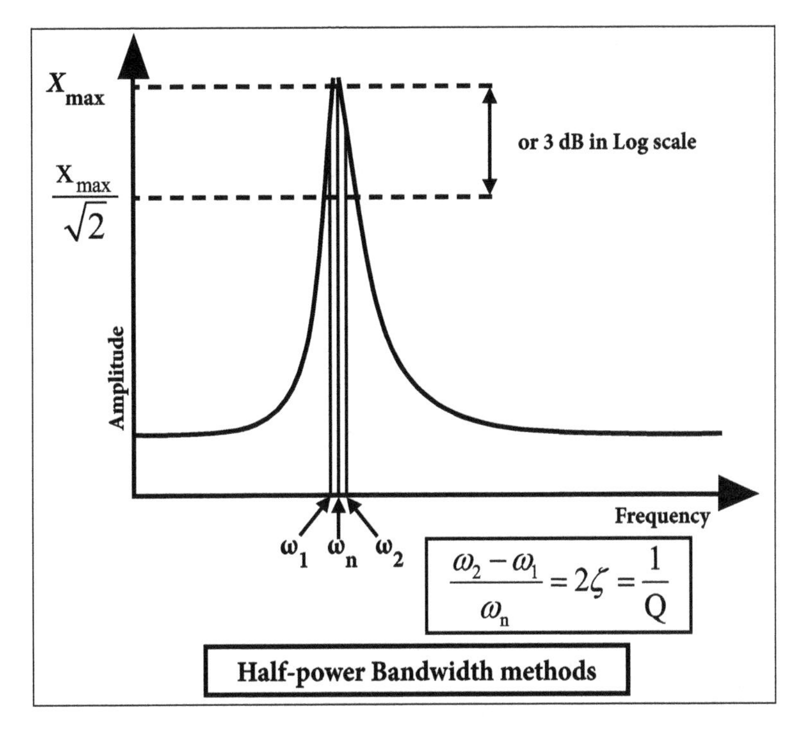

Logarithmic decrement method is used to measure damping in time domain. In this method, free vibration displacement amplitude history of a system to an impulse is measured and recorded. A typical free decay curve is shown .Logarithmic decrement is the natural logarithmic value of the ratio of two adjacent peak values of the displacement in free decay vibration.

To estimate damping ratio from frequency domain, we may use the half-power bandwidth method. In this method, FRF amplitude of the system is obtained first. Corresponding to each natural frequency, there is a peak in the FRF amplitude. 3 dB down from the peak there are two point corresponding to half power point, as shown in figure below. The more the damping, the more the frequency ranges between this two point. Half-power bandwidth BD is defined as ratio of the frequency range between the two half power points to the natural frequency at this mode.

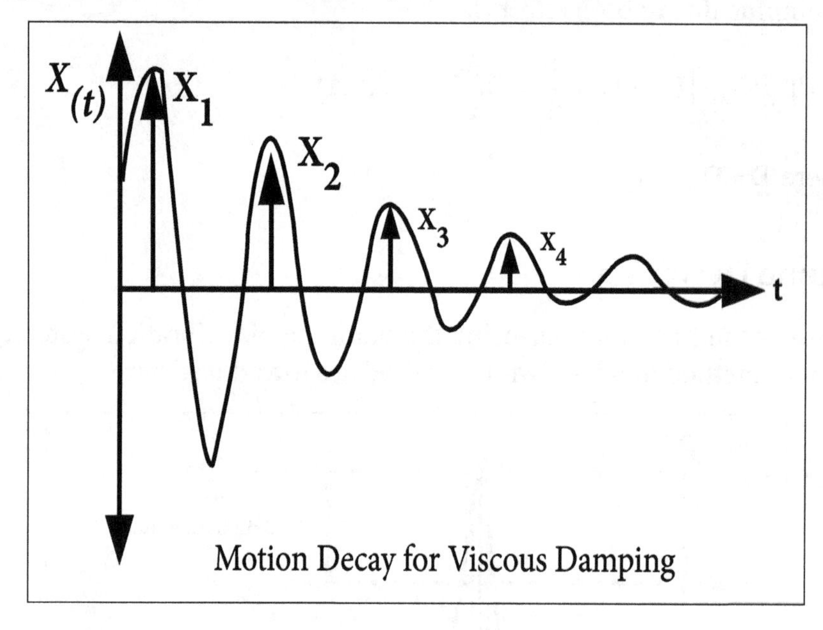

Motion Decay for Viscous Damping

$$\delta = \frac{2\pi\zeta}{\sqrt{1-\zeta^2}} \quad \delta = \frac{1}{n}\ln\left|\frac{x_1}{x_{a+1}}\right| \text{ and } \zeta = \frac{\zeta}{\sqrt{4\pi^2+\zeta^2}}$$

Calibration of the Ballistic Galvanometer

For small deflections of the ballistic galvanometer the charge passing through galvanometer is proportional to the deflection, d, which is:

$$\Delta q = kd$$

In order to find the constant of the galvanometer, k, known amount of charge should be used and deflection should be measured.

One can obtain the known amount of charge by discharging a capacitor of known capacitance, C, though the galvanometer. To do that, build an electric circuit shown in the figure below.

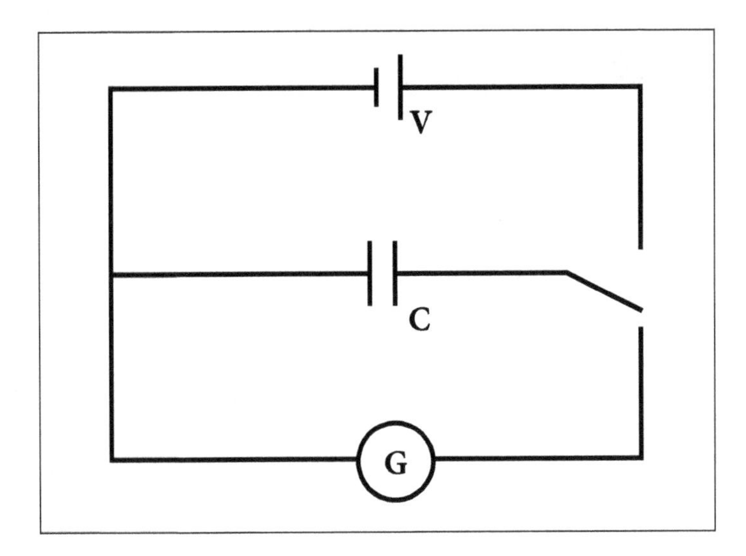

First close key in the upper position to charge the capacitor from the power supply of the known potential difference, V. Then move the key to lower position to discharge the capacitor through the galvanometer and measure the deflection of galvanometer, d. (Once again do it twice switching polarity and then take the average of the two readings). In this case the charge is $\Delta q = CV$, so:

$$CV = kd$$

Collect several data points for different values of C and V. Think what needs to be graphed in order to find k.

Flux Meter

Flux meter is a special type of ballistic galvanometer in which the controlling torque is very small and electromagnetic damping is very heavy.

Construction

The construction of a flux meter is shown in Figure (a). It consists of a coil C of small cross-section suspended from a spring support S by means of a single silk thread T and hangs with its parallel sides in the narrow air-gap of a permanent magnet system. The coil moves in this narrow air-gap. The current is led into the coil C by spirals SP of very thin, annealed silver strips. This reduces the controlling torque to minimum. Flux meter is fitted with a pointer attached to the moving system and a scale. It is graduated in terms of flux-turns. Since, this flux meter was designed by Grassot hence it is also referred as Grassot flux meter.

In modern flux meters the coil is fitted with pivots and mounted in jewel-led bearings, the current being led into the coil by fine ligaments. This form of construction is more robust than suspended instruments.

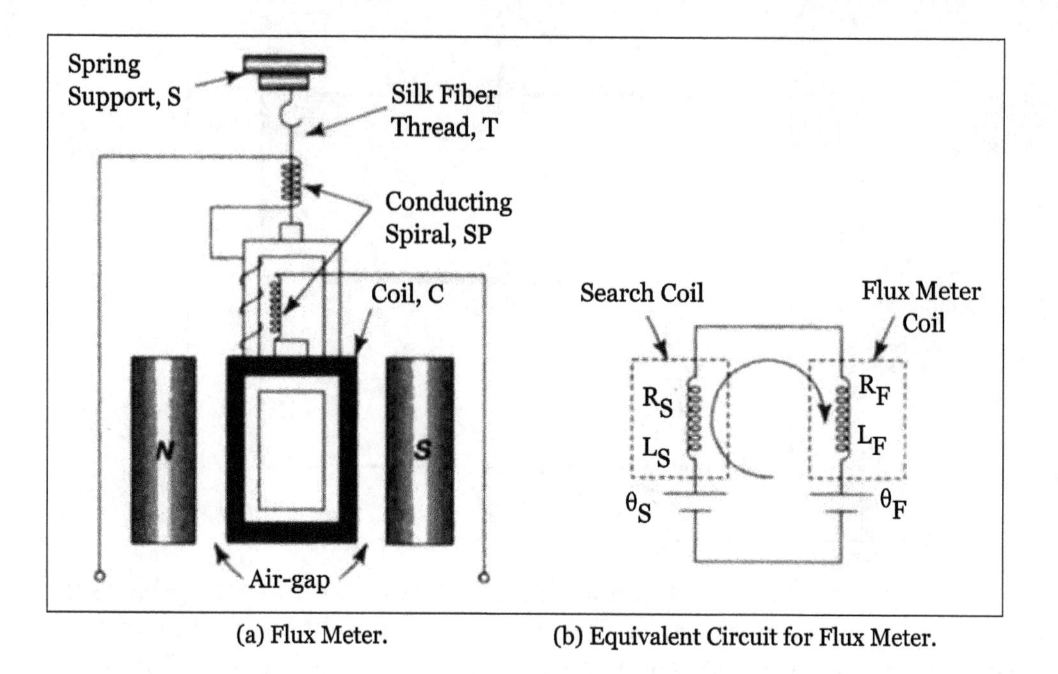

(a) Flux Meter. (b) Equivalent Circuit for Flux Meter.

Working Principle

Assume that the controlling torque is negligibly small and also that the air damping and friction are negligible. In this case, the flux meter would remain in the deflected position indefinitely. Actually the pointer returns very slowly to zero, but readings may be taken by observing difference in deflection at the beginning and end of the change in flux to be measured without waiting for the pointer to return to zero, the scale being uniform. As shown in Figure (b), the resistance R_S of the search coil circuit connected to the flux meter should be fairly small because variation in this resistance of several ohms usually have a negligible effect upon deflection.

If Φ_1 and Φ_2 are the inter-linking fluxes at the beginning and at end of the change in flux to be measured and θ_1 and θ_2 are the corresponding deflection, then:

$$=\left[\frac{b(R_F+R_S)}{G}+G\right](\theta_2-\theta_1)=N(\Phi_2-\Phi_1) \qquad ....(1)$$

Where:

b = the damping constant.

R_F = the resistance of flux meter.

R_S = the resistance of search coil.

G = displacement constant of flux meter.

N = number of turns on the search coil connected to flux meter.

If the total circuit resistance ($R_F + R_S$) is kept small and damping constant b is small, then the Equation (1) is simplified as:

$$G(\theta_2 - \theta_1) = N(\Phi_2 - \Phi_1)$$

If $d\theta$ the change in deflection and $d\Phi$ is the change in flux, then:

$$G\,d\theta = N\,d\Phi$$

$$d\theta = \frac{N}{G}d\Phi$$

Advantages:

- The flux meter is portable.
- It has advantage over galvanometer because the length of time taken for the change in flux producing the deflection need not be small. The deflection obtained for a given change of flux inter-linking with the search coil connected to the flux meter will be same whether the time taken for the change be a fraction of a second or as much as one or two minutes.
- Flux meter is widely used in place of the ballistic galvanometer except for precise laboratory measurements.

Disadvantages:

- The sensitivity of flux meter is inferior to the ballistic galvanometer.
- Due to heavy damping, the pointer returns to zero very slowly.

2.2 Ammeter and Voltmeter

S. No.	Analog Meters	Digital Meters
1.	Errors such as human reading error, interpolation error and parallel error are more.	Due to digital display such errors are reduced.
2.	They have limited range and cannot be varied.	The range can be varied easily.
3.	Lesser accuracy and precision.	Greater accuracy and precision.
4.	Speed of reading is slow.	Speed of reading is fast.
5.	These meters were more affected by noise.	They are less affected by noise.

6.	They cannot be programmed.	They can be programmed.
7.	The output cannot be recorded.	The output can be recorded.

Digital Voltmeters

Types of DVMs:

Ramp Type Digital Voltmeter

The operating principle of a ramp type digital voltmeter is to massive the time that a linear ramp voltage takes to change from the level of input voltage to zero voltage (or vice versa). This time interval is measured with an electronic time interval counter and count is displayed as a number of digits on electronic indicating tubes of the output readout of the voltmeter.

The conversion of a voltage value of a time interval is shown in the timing diagram.

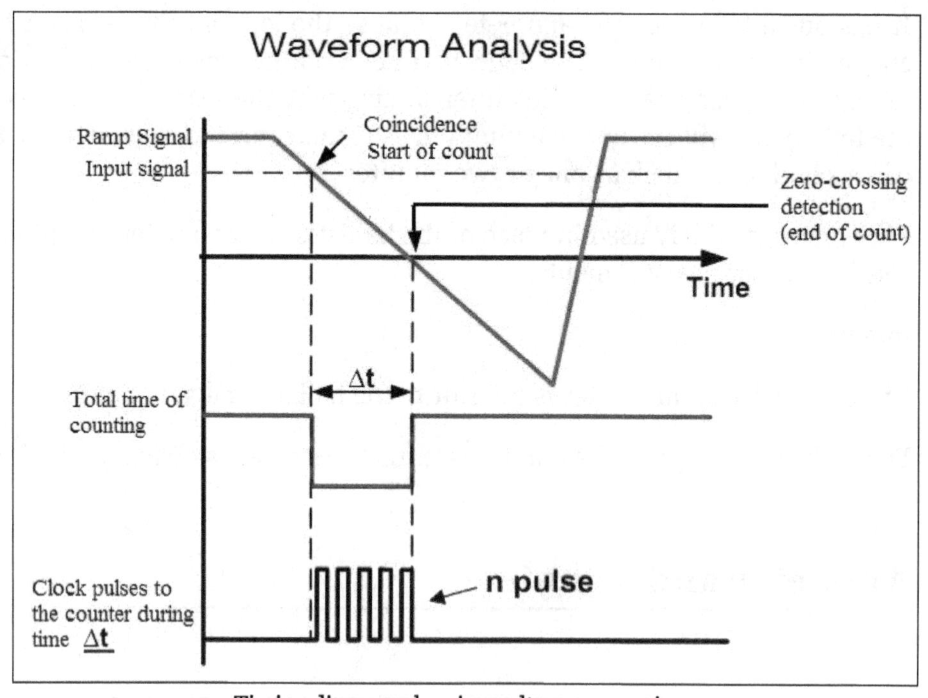

Timing diagram showing voltage conversion.

At the start of measurement, a ramp voltage is initiated. A negative ramp is shown in figure but a positive going ramp may also be used. The ramp voltage value is continuously compared with the voltage being measured. At the instant, the value of ramp voltage is equal to that of unknown voltage a coincidence circuit called an input comparator, generates a pulse which opens a gate. The ramp voltage continue to decrease till it reaches ground level. At this instant another comparator called ground comparator generates a pulse and closes the gate.

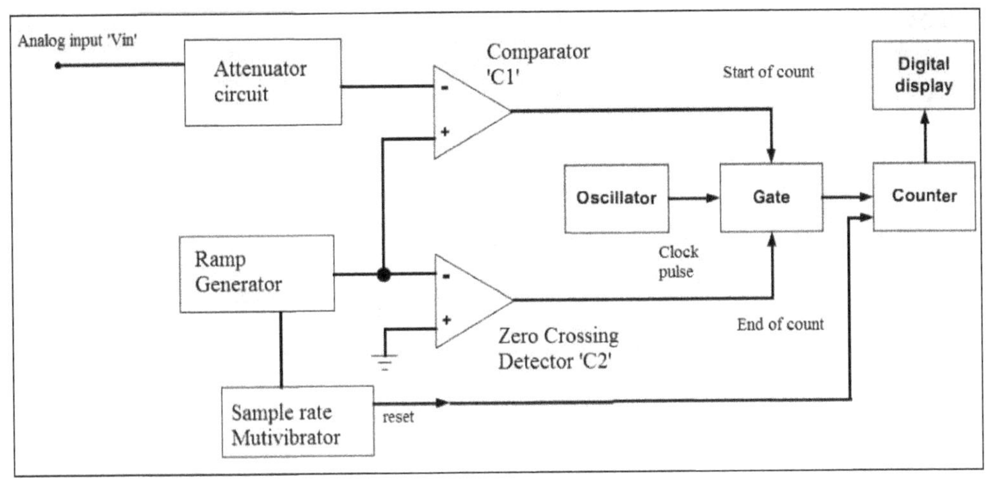

Block diagram of a ramp DVM.

The time elapsed between opening and closing of the gate is 't' as indicated in the above figure. During this time interval, pulses from a clock pulse generator pass through the gate and are counted and displayed.

The decimal number as indicated by the readout is measure of the value of input voltage.

The sample rate multi vibrator determine the rate at which the measurement cycles are initiated. The sample rate circuit provides an initiating pulse for the ramp generator to start its next ramp voltage. At the same time, it sends a pulse to the counters which sets all of them to 0. This momentarily removes the digital display of the readout.

Dual Slope Integrating type Digital Voltmeter

In this type, the most popular method of analog to digital conversion is used. The basic principle is that the input signal is integrated for fixed interval and then same integrator is used to integrate the reference voltage with reverse slope.

When the switch S_1 is in position 1, the capacitor starts charging from zero. The rate of charging is proportional to the input voltage level. The output of the op-amp is given by:

$$V_{out} = \frac{1}{R_1 C} \int^{V_1} V_{in}\, dv$$

$$V_{out} = \frac{V_{in}}{R_1} \frac{l_1}{C} \qquad ...(1)$$

After a time interval, the input voltage is disconnected and a negative voltage $-V_{net}$ is connected by throwing the switch in position 2. In this position, the output of op-amp is given by:

$$V_{out} = \frac{}{R\,C} \int -V_{ref}\, dv$$

$$V_{out} = \frac{-V_{ref}\, l_2}{R_1\, C} \qquad ...(2)$$

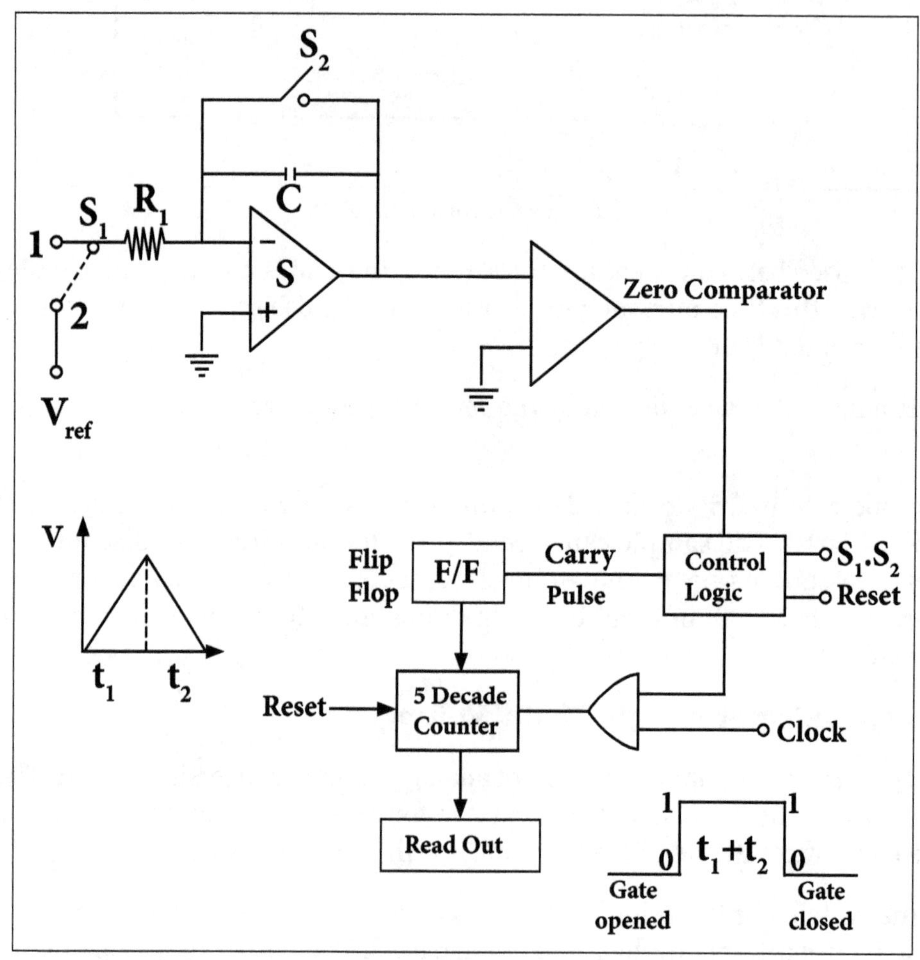

Block diagram of Dual slope integrating type digital voltmeter.

Subtracting (1) from (2):

$$V_{out} - V_{out} = 0 = \frac{V_{ref}\, l_2}{R_1\, C}\left(\frac{V_{in}\, l_2}{R_1\, C}\right)$$

$$\frac{V_{ref}\, l_2}{R_1\, C} = \frac{V_{in}\, l_1}{R_1\, C}$$

$$V_{in} = V_{ref}\left(\frac{1_2}{1_1}\right)$$

The input voltage depends on the time period t_1 and t_2 not on the values of R_1 and C.

At the start of the measurement, the counter is reset to zero and the flip-flop is also zero. This is given to control logic and this control sends a signal to close switch to position 1 and integration of input voltage starts.

It continues till the time period t_1 as the output of the integrator changes from its zero value, the zero comparator output changes its state. This provides a signal to control logic which in turns opens the gate and the counting of lock pulse starts.

If the counter exceeds 9999, then the flip-flop output gets activated to logic level 1. This changes the switch position S_1 to S_2. Then - V_{ref} is gets connected to op-amp which gives negative slope.

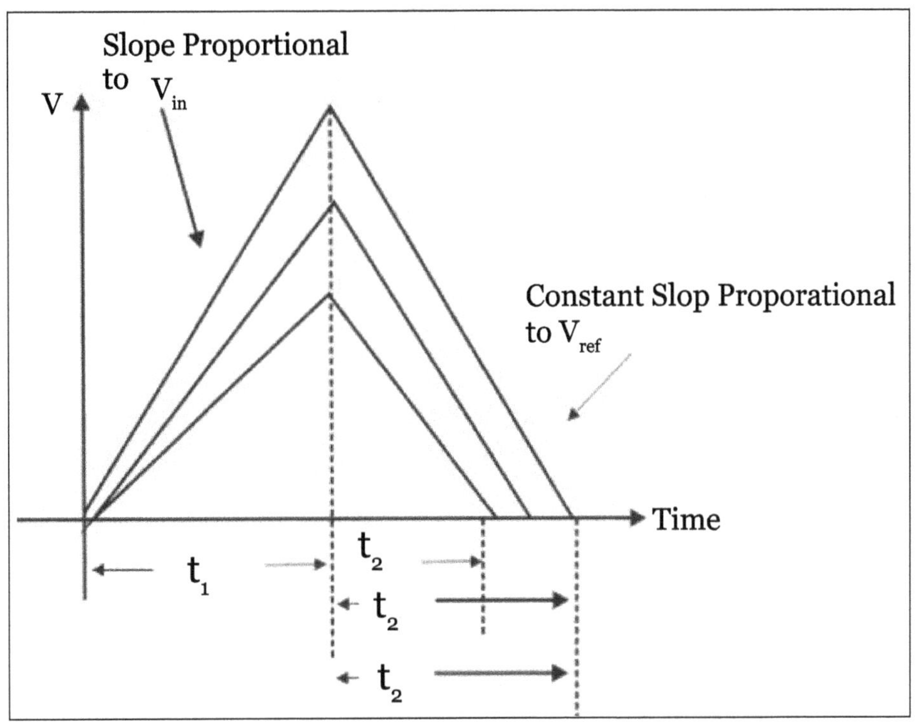

Basic principle of dual slope method.

The output decreases linearly and attains zero value, when the capacitor C gets fully discharged. The gate closed after time t_2 the counting operation stops and transferred to the readout.

$$V_{in} = V_{ref}\frac{n_2}{n_1}.$$

Successive Approximation Type ADC

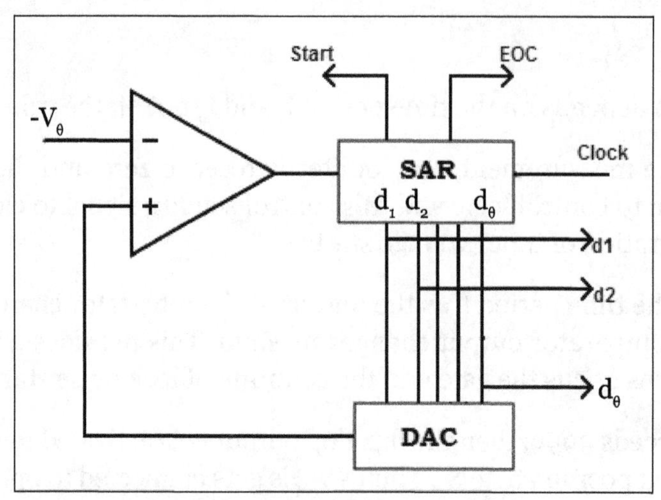

Successive approximation type ADC.

The figure shows an 8-bit converter which requires 8 clock pulses to obtain digital-output. It consists of successive approximation register (SAR), operational amplifier and DAC. SAR is used to find the required value of each bit by trial and error.

When start command is given, SAR sets MSB, d_1 = 1 with all other bits to zero, so that the trial ends is 1000 000. The output from DAC, V_d is compared with analog input V_a in comparator.

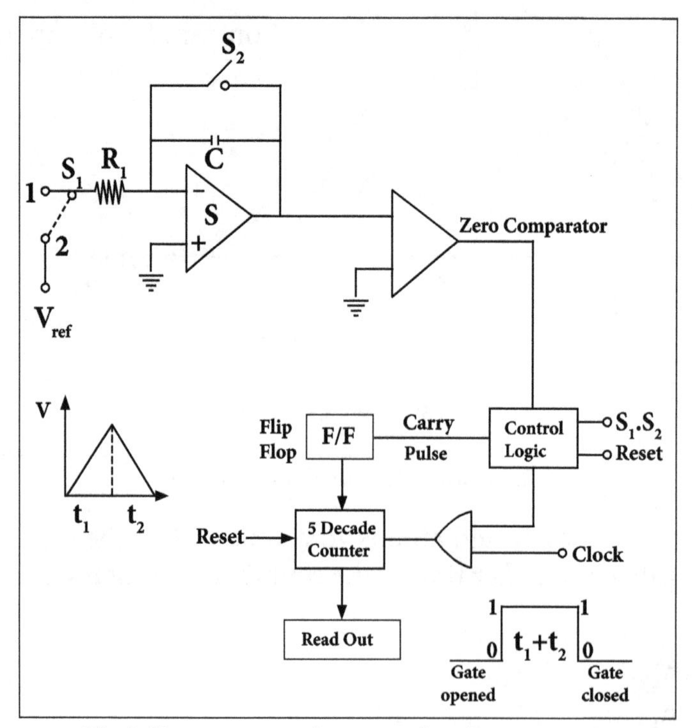

Block diagram successive approximation type ADC.

If $V_a > V_d$, then 1000 0000 is less than correct digital representation.

If $V_a < V_d$, then 1000 0000 is greater than correct digital representation.

Now MSB is reset to 0 and goes on to the next lower significant bit. This procedure will be separate of for all subsequent bits (i.e., from MSB to LSB), one at a time, until all bit position have been tested. When DAC crosses V_a then the comparator change its state and this is end of conversion (EOC). It has high resolution's high speed.

Moving Iron (MI) Instruments

One of the most accurate instruments used for both AC and DC measurement is moving iron instrument.

There are two types of moving iron instrument:

Attraction Type M.I. Instrument

Construction

The moving iron fixed to the spindle is kept near the hollow fixed coil. The pointer and balance weight are attached to the spindle, which is supported with jeweled bearing. Here air friction damping is used.

Principle of Operation

The current to be measured is passed through the fixed coil. As the current is made to flow through the fixed coil, a magnetic field is produced. By magnetic induction the moving iron gets magnetized. The north pole of moving coil is attracted by the south pole of fixed coil. Thus the deflecting force is produced due to force of attraction. Since the moving iron is attached with the spindle, the spindle rotates and the pointer moves over the calibrated scale. But the force of attraction depends on the current flowing through the coil.

Torque Developed by M.I

Let 'θ' be the deflection corresponding to a current of 'i' amp Let the current increases by di, the corresponding deflection is '$\theta + d\theta$'.

There is change in inductance since the position of moving iron change w.r.t the fixed electromagnets. Let the new inductance value be the current change by 'di' is dt seconds. Let the emf induced in the coil be 'e' volt.

Eqn. gives the energy is used in to two forms. Pan of energy is stored in the inductance. Remaining energy is converted in to mechanical energy which produces deflection.

$$e = \frac{d}{dt}(Li) = L\frac{di}{dt} + i\frac{dL}{dt}$$

Multiplying by 'idt' in equestion ...(1)

$$e \times idt = L\frac{di}{dt} \times idt + i\frac{dL}{dt} \times idt$$

$$e \times idt = Lidi + i^2 dL$$

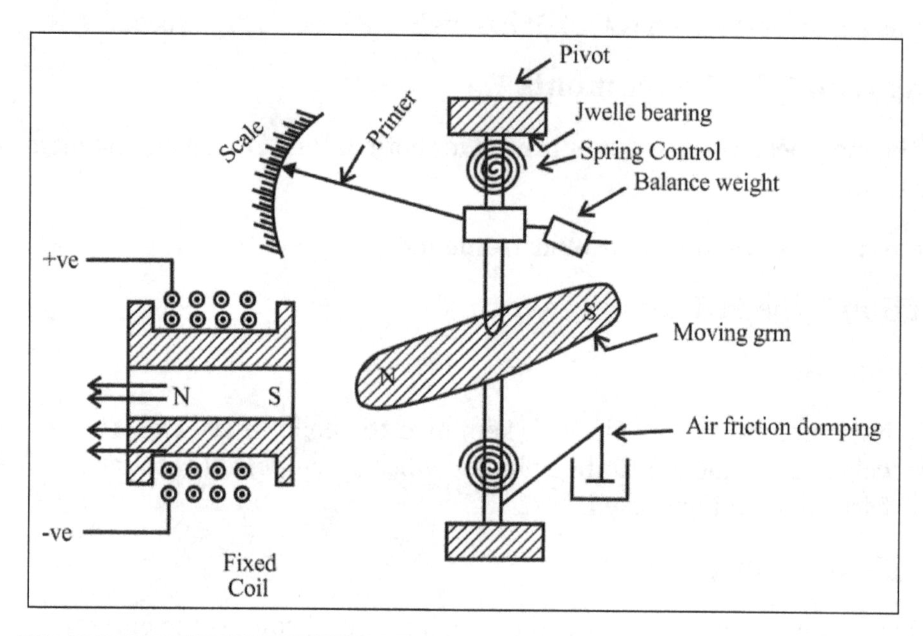

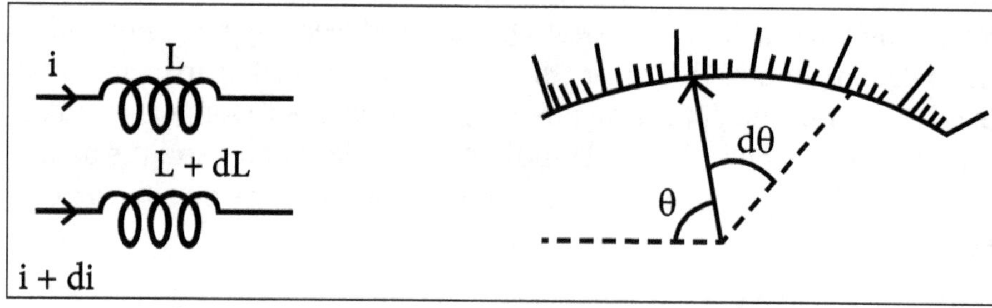

Change in energy stored=Final energy-initial energy stored:

$$= \frac{1}{2}(L + dL)(i + di)^2 - \frac{1}{2}Li^2$$

$$= \frac{1}{2}\left\{(L + dL)\left(i^2 + di^2 + 2idi\right) - Li^2\right\}$$

$$= \frac{1}{2}\left\{(L + dL)\left(i^2 + 2idi\right) - Li^2\right\}$$

$$= \frac{1}{2}\left\{Li^2 + 2Lidi + i^2 dL + 2ididL - Li^2\right\}$$

$$=\frac{1}{2}\left\{2Li\,di+i^2\,dL\right\}$$

$$=Li\,di+\frac{1}{2}i^2\,dL \qquad ...(2)$$

Mechanical work to move the pointer by $d\theta$:

$$=T_d\,d\theta \qquad ...(3)$$

By law of conservation of energy:

Electrical energy supplied = Increase in strored energy = mechanical work done.

Input energy = Energy storege + mechanical energy

$$Li\,di+i^2\,dL=Li\,di+\frac{1}{2}i^2\,dL+T_d\,d\theta \qquad ...(4)$$

$$\frac{1}{2}i^2\,dL=T_d\,d\theta \qquad ...(5)$$

$$T_d=\frac{1}{2}i^2\frac{dL}{d\theta} \qquad ...(6)$$

At steady state condition,

$$T_d=T_c$$

$$\frac{1}{2}i^2\frac{dL}{d\theta}=K\theta \qquad ...(7)$$

$$\theta=\frac{1}{2K}i^2\frac{dL}{d\theta} \qquad ...(8)$$

$$\theta\propto i^2 \qquad ....(9)$$

When the instruments measure AC, $\theta \propto i^2_{rms}$

Scale of the instrument is non-uniform.

Advantages:

- MI can be used in AC and DC.

- It is cheap.
- Supply is given to a fixed coil, not in moving coil.
- Simple construction.
- Less friction error.

Disadvantages:

- It suffers from eddy current and hysteresis error.
- Scale is not uniform.
- It consumes more power.
- Calibration is different for AC and DC operation.
- Repulsion type moving iron instrument.

Construction

The repulsion type instrument has a hollow fixed iron attached to it. The moving iron is connected to the spindle. The pointer is also attached to the spindle and supported with jeweled bearing.

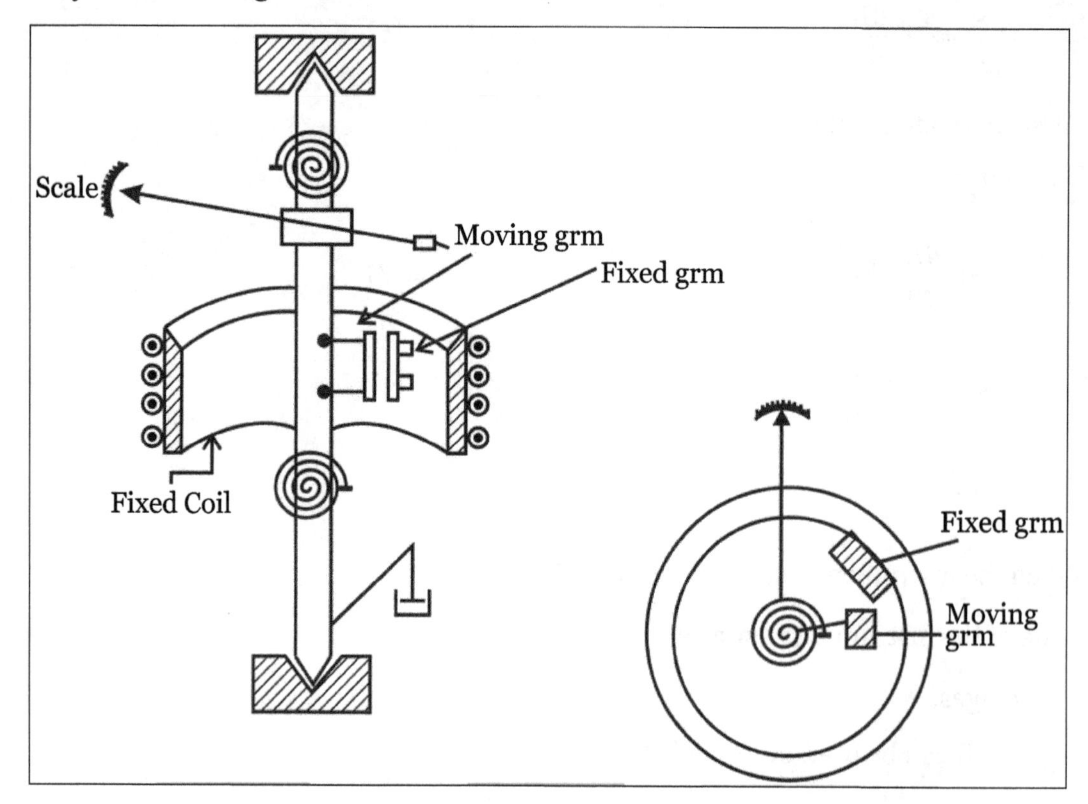

Principle of Operation

When the current flows through the coil, a magnetic field is produced by it. So both fixed iron and moving iron are magnetized with the same polarity, since they are kept in the same magnetic field. Similar poles of fixed and moving iron get repelled. Thus the deflecting torque is produced due to magnetic repulsion. Since moving iron is attached to spindle, the spindle will move. So that pointer moves over the calibrated scale.

Damping: Air friction damping is used to reduce the oscillation.

Control: Spring control is used.

Ammeters and Multimeters

An ammeter is an instrument for measuring the electric current in amperes in a branch of an electric circuit. It must be placed in series with the measured branch and must have very low resistance in order to avoid significant alteration of the current it has to measure.

By contrast, a voltmeter should be connected in parallel. The analogy with an inline flow meter in a water circuit can help visualize why an ammeter must have low resistance and why connecting an ammeter in parallel can damage the meter. Modern solid-state meters have digital readouts, but principles of operation can be better appreciated by examining the older moving coil meters based on galvanometer sensors.

Ammeters are connected in series in the circuit whose current is to be measured. The power loss in an ammeter is IRa. Therefore, ammeters should have a low electrical resistance so that they have a small voltage drop and consequently absorb small power.

Four types of analog ammeter used for instrumentation are:

Permanent Magnet Moving Coil Instrument

The permanent magnet moving coil instruments are most accurate type for DC measurements. The action of these instruments is based on the motoring principle.

When a current carrying coil is placed in the magnetic field produced by permanent magnet, the coil experiences a force and moves. As the coil is moving and magnet is permanent, the instrument is called permanent magnet moving coil instrument. The basic principle is called D' Arsonral principle. The amount of force experienced by the coil is proportional to the current passing through the coil.

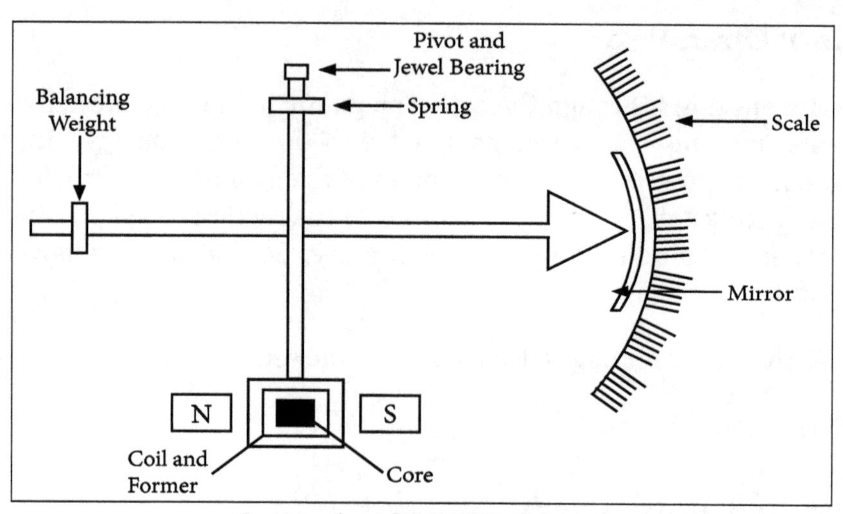

Construction of PMMC Instrument.

The PMMC instrument is shown in the figure. The moving coil is either rectangular or circular in shape. It has a number of turns of fine wire. The coil is suspended so that it is free to turn about vertical axis. The coil is placed in uniform, horizontal and radial magnetic field of a permanent magnet in the shape of horse-shoe. The iron core is spherical if coil is circular and is cylindrical if the coil is rectangular. Due to iron core, deflecting torque increases, increasing the sensitivity of the instrument.

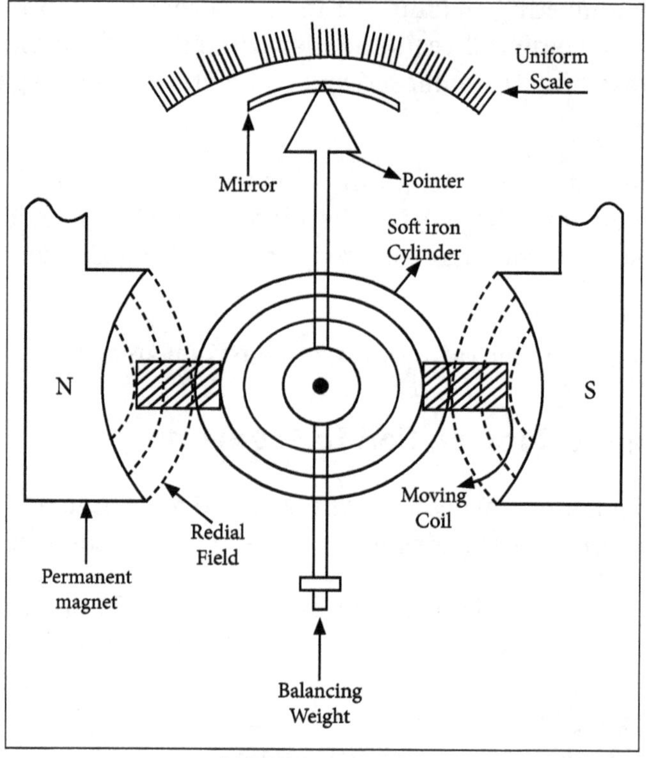

PMMC Instrument.

The controlling torque is provided by two phosphor bronze hair springs. The damping torque is provided by eddy current damping. It is obtained by movement of the aluminum former, moving in the magnetic field of the permanent magnet.

The pointer is carried by spindle and it moves over a graduated scale. The pointer has light weight so that it can deflect rapidly. The mirror is placed below the pointer to get accurate reading by removing the parallax. The weight of the instrument is normally counter balanced by the weights situated diametrical opposite and rigidly connected to it. The scale markings of the basic DC PMMC instruments are usually linearly spaced as the deflecting torque and hence the pointer deflections are directly proportional to the current passing through the coil.

The top view of PMMC instrument is shown in the figure. In a practical PMMC instrument, a Y-shaped member is attached to the fixed end of the front control spring. An eccentric pin through instrument case engages the Y-shaped member, so that the zero position of the pointer can be adjusted from outside.

Torque Equation

The equation for the developed torque can be obtained from the basic law of the electromagnetic torque. The deflecting torque is given by:

Therefore,

$$T_d = NBA\,I$$

Where,

T_d = Deflecting torque in N-m.

B = Flux density in air gap, Wb/m^2.

N = Number of turns of the coil.

A = Effective coil area, m^2.

I = Current in the moving coil, Ampere.

$$T_d = GI$$

Where,

G = NBA = Constant.

The controlling torque is provided by the springs and is proportional to the angular deflection of the pointer.

$$T_c = K\theta,$$

Where,

T_c = Controlling torque.

K = Spring constant, Nm/rad or Nm/deg.

θ = Angular deflection.

For the final steady state position:

$$T_d = T_c \text{ i.e., } GI = K\theta.$$

Therefore,

$$\theta = \left[\frac{G}{K}\right] I \quad \text{or} \quad I = \left(\frac{K}{G}\right)\theta.$$

The pointer deflection can therefore be used to measure current.

As the direction of the current through to the coil changes, the direction of the deflection of the pointer also changes. Hence such instruments are well suited for the DC measurements.

Digital Multimeter

The digital multimeter is an instrument which is capable of measuring voltages, DC voltages, AC and DC currents and resistances over several ranges.

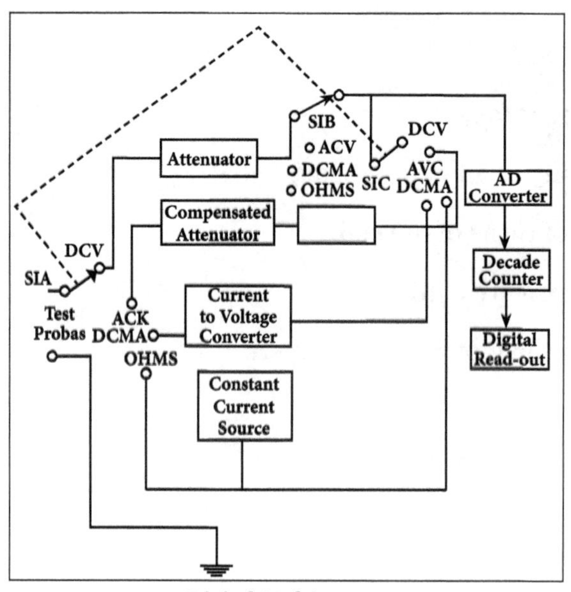

Digital Multimeter.

The current is converted to voltage by passing it through a low shunt resistance. The AC quantities are converted to DC by employing various rectifiers and filtering circuits.

For resistance measurements, the meter consists of a precision low current square that is applied across the unknown resistance which gives DC voltage.

All the quantities are digitized using analog to digital converter and displayed in the digital form on the display. In addition, the output of digital multimeters can also be used to interface with some other equipment.

Advantages:

- The accuracy is very high.
- The input impedance is very high.
- The size is compact and cost is also low.

Problem

A PMMC ammeter gives reading of 40 mA when connected across two opposite corners of a bridge rectifier, the other two corners of which are connected in series with a capacitor to 100 k, 50 Hz supply. Let us determine the capacitance.

Solution:

Given data:

$$i_{av} = 40 \text{ mA}$$

$$V_m = 100 \text{ k}$$

$$f = 50 \text{ Hz}$$

Formula to be used:

$$V_m = \frac{i_{av}}{4C_f}$$

$$\text{W.K.T}, V_m = \frac{i_{av}}{4C_f}$$

$$= C = \frac{i_{av}}{V_m 4_f} = \frac{40 \times 10^{-3}}{100 \times 10^3 \times 4 \times 50}$$

$$= 2 \times 10^{-9} \text{F}.$$

2.3 Potentiometer

Measurement of Resistance using DC Potentiometer

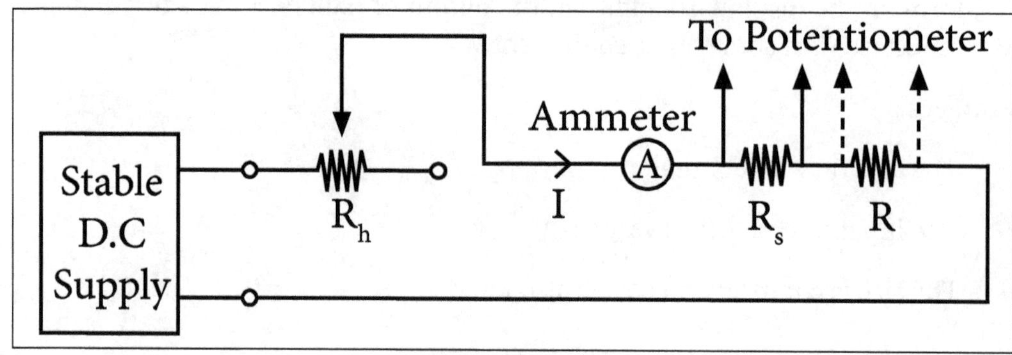

DC potentiometer.

A resistor whose resistance is to be measured is connected in series with a standard resistor of resistance R_s. The current through the circuit is supplied by stable D.C supply and it is controlled by a rheostat R_h.

Due to the current I, voltages are developed across R_s and R. Both are then measured by using a DC potentiometer.

Voltage across standard resistance:

$$V_{RS} \quad IR_S$$

Voltage across unknown resistance be:

$$V_R = IR.$$

$$\frac{VR}{V_{RS}} = \frac{R}{R_S}$$

$$R_S = R\left(\frac{V_R}{V_{RS}}\right) \rightarrow \text{unkown resistance.}$$

Duo-Range DC Potentiometer

By keeping direct reading feature, the basic potentiometer may be modified to add a second range with usually a second factor such as 0.01. Such a potentiometer is called duo-range potentiometer.

The duo-range potentiometer is the modified version of Crompton's DC potentiometer in which additional range selector switch is used as shown in the figure.

The working battery B of sufficient capacity is connected in series with two variable resistances R_1 and R_2 which are used for regulating the current through potentiometer during standardization.

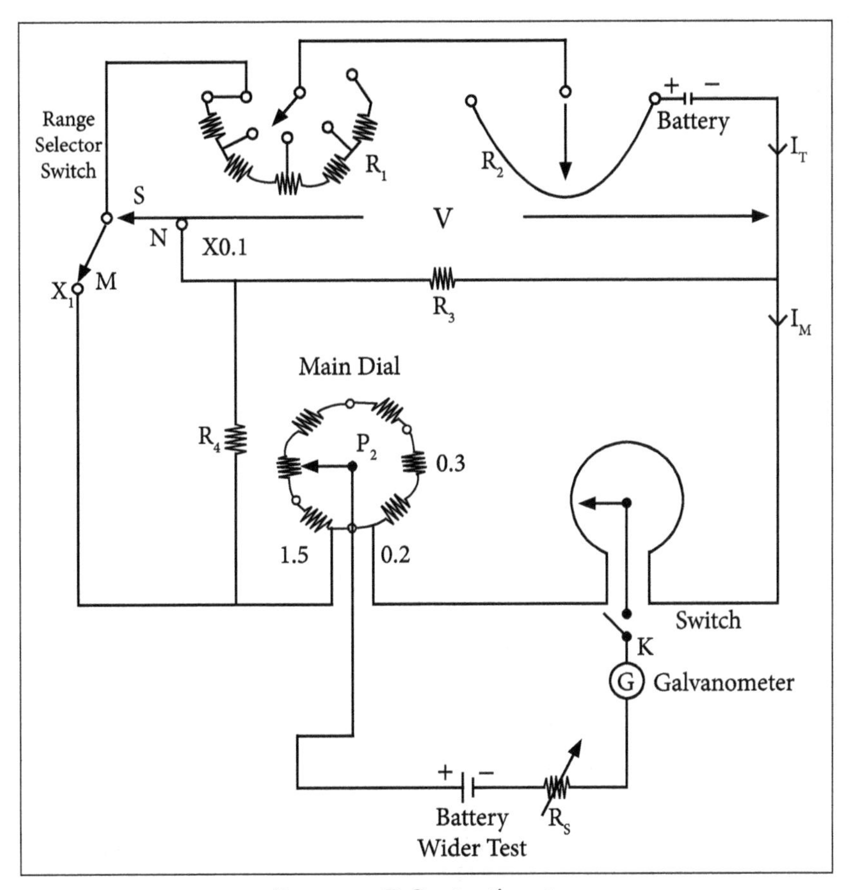

Due range D.C potentiometer.

Let the total resistance of measuring circuit including the main dial and slide wire be RM, where RM is the addition of the main dial resistance and slide wire resistance. The slide wise resistance is 10 Ω. The resistance of each coil in the main dial is same as that of the slide wire i.e., 10 Ω and there are 15 steps and so the total main dial resistance is 150 Ω. Let the current through the measuring circuit be 10 mA. Hence the total drop across R_M is given by:

$$VRM = I_M \times R_m = (10\times10^{-3})(160) = 1.6\,V \ldots(1)$$

The second range can be obtained by changing position of switch S from M to NX, to X0.1 range. When this range is selected, the current I_M must reduce to $1/10^{th}$ of its original value. When the switch S is at position M, the current adjusted equals 10 mA. So when switch S is moved to position N, I_M must reduce to 1 mA in order to products voltage drop of 0.16 V across R_M.

It is very important to design a circuit in such a way that the instrument should enable the user to change the measuring range without standardizing the instrument again and again either by adjusting R_1 and R_2 or by changing the voltage of working battery B. In other words, once the instrument is standardized for XI range, there is no need to standardize again for X0.1 range. This condition can be fulfilled only if the voltage V remains same for both the positions of range selector switch S. And this is possible only if the total battery current I_T is same for both the ranges.

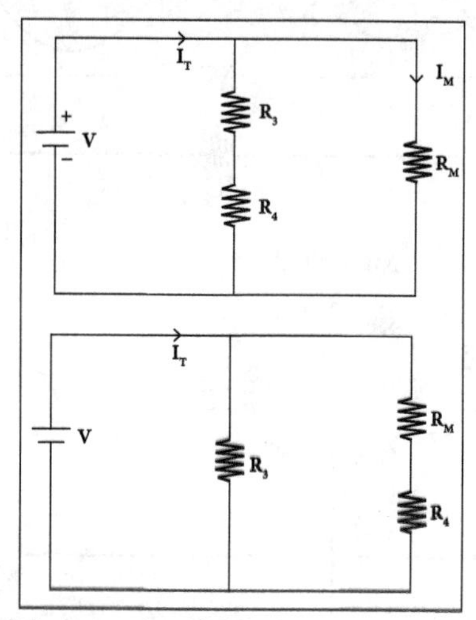

Simplified circuits of duo-range potentiometer.

The operation of duo-range potentiometer can be explained in simple form as shown. First consider that the range selector switch S is at position M i.e., on range of XI. Then the total measuring resistance R_M gets shunted by series combination of range resistors R_3 and R_4 as shown in figure .The current through R_M is I_M while total current is I_T. Now consider that the range selector switch S is moved to position N i.e., on range of X0.1. The range resistor R_3 shunts the series combination of total measuring resistance R_M and range resistor R_4 as shown in figure Now the current flowing through the branch consisting R_M and R_H is I_M' which is (0.1) I_M and still total current is I_T.

In order to have the same current I_T for both ranges, the condition is:

$$(R_3+R_4)\| R_M = R_3 \| (R_M+R_4)$$

$$\therefore \frac{(R_3+R_4)R_M}{R_3+R_4+R_M} = \frac{R_3(R_M+R_4)}{R_3+R_4+R_M}$$

$$\therefore R_3 R_M + R_4 R_M = R_3 R_M + R_3 R_4$$

$$\Rightarrow R_M = R_3$$

Above equation indicates that range resistance R_3 must be selected same as total measuring resistance R_M so as to keep total current, supplied by battery B, same for both the ranges.

Now the second condition states that the current I_M' when switch S is at position V must be equal to 0.1 I_M, where I_M is current through R_M when switch S is at position M.

$$\therefore \quad I_M = 0.1 I_M \qquad \therefore \frac{V}{R_M + R_4} = 0.1\left[\frac{V}{R_M}\right]$$

$$\therefore \quad R_M = 0.1(R_M + R_4) \Rightarrow R_4 = 9R_M = 9R_3$$

Thus by properly designing values of R_3 and R_4 we can achieve high revolution in measurement using duo-range potentiometers.

Gall-Tinsley (Co-Ordinate Type) AC Potentiometer

This potentiometer consists of two separate potentiometer circuits enclosed in a common case. One is called the "in-phase" potentiometer and the other is called the "quadrature" potentiometer. The slide-wire circuits are supplied with currents which have a phase difference of 90°. Since the two slide-wire currents are in quadrature, the two measured values are the quadrature components of the unknown voltage. If these measured values are V_1 and V_2 respectively, then the unknown voltage is given by $V = \sqrt{V_1^2 + V_2^2}$ and its phase difference from the current in the "in-phase" potentiometer slide-wire circuit is given by the angle θ where tan $\theta = V_2 / V_1$.

Figure shows the connections of the potentiometer, simplified somewhat for the purpose of clearness. The in-phase and quadrature potentiometer circuit are shown, with their sliding contacts bb' and cc' and rheostats Rand R' for current adjustment. The supplies to the potentiometer are obtained from a single-phase supply by means of the arrangement shown in figure.

T_1 and T_2 are two step-down transformers which supply about 6 V to the potentiometer circuits. They also serve to isolate the potentiometer from the line and are usually provided with earthed screens between the windings. The supply to T_2 is obtained through a variable resistor R and variable capacitor C for the purpose of phase splitting. Quadrature phase displacement is obtained by adjusting C and R.

Referring again to figure, V.G. is a vibration galvanometer (tuned to the supply frequency) and K is its key. A is a reflecting dynamometer instrument for maintaining the current in the two slide-wires at the standard value (50 mA). S_1 and S_2, are two "sign-changing" switches which may be necessary to reverse the direction of the unknown emf applied to the slide-wires.

The necessity of these switches depends on the relative phases of the unknown and slide wire-voltages. S_3 is a selected switch by which the unknown voltages to be measured are placed in the circuit. There are four pairs of terminals for the application of such voltages, the connection to only one pair to which an unknown voltage V is applied being shown in the figure. The selector switch, when in the position shown in the figure called the "test position" allows the current in the quadrature potentiometer slide-wire to be compared with that in the in-phase potentiometer wire, utilizing the mutual inductance M for the purpose.

The current in the in-phase potentiometer wire is first adjusted to its standard value by means of a direct current supply and a standard cell, the vibration galvanometer is replaced by a galvanometer of the d 'Arsenval type for this purpose. The dynamometer ammeter is of the torsion head type and the torsion head is turned to give zero deflection on direct current. This setting is left untouched during the calibration value of unknown voltage now becomes a matter for evaluating alternating current, the slide-wire current being adjusted to give zero defection again. The vibration galvanometer is then placed in current and the direct current supply is replaced by the alternating supplies.

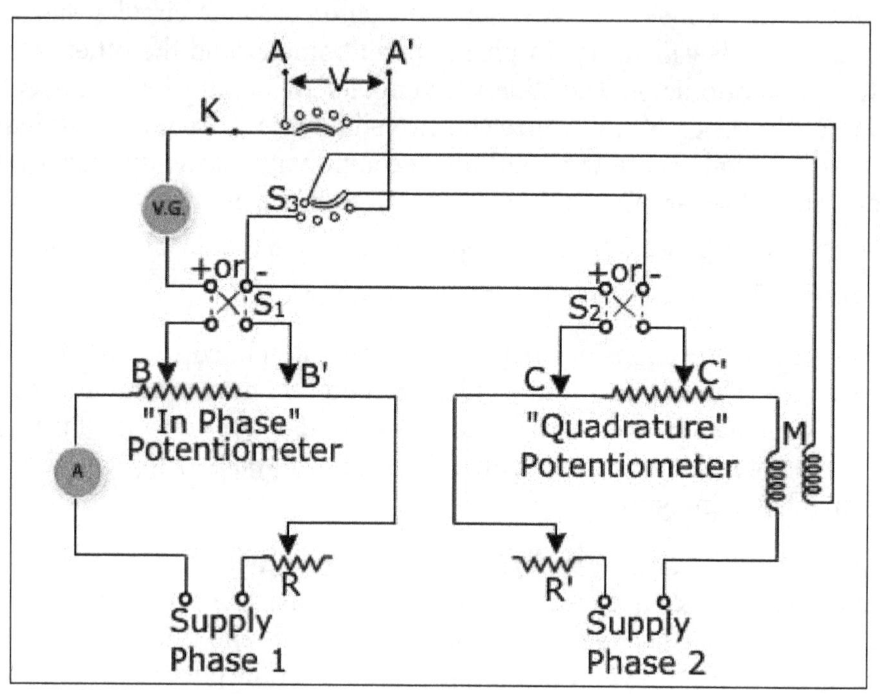

Connections of gall-Tinsley potentiometer.

Now, the magnitude of the current in the quadrature potentiometer wire must be the same as that in the in-phase potentiometer namely, the standard value of 50 mA. These two currents must also be exactly in quadrature. Rheostat R is adjusted until the current in the in-phase potentiometer wire is the standard value (as indicated on A). The selector switch S_3 is then switched on to the test position (shown in figure).

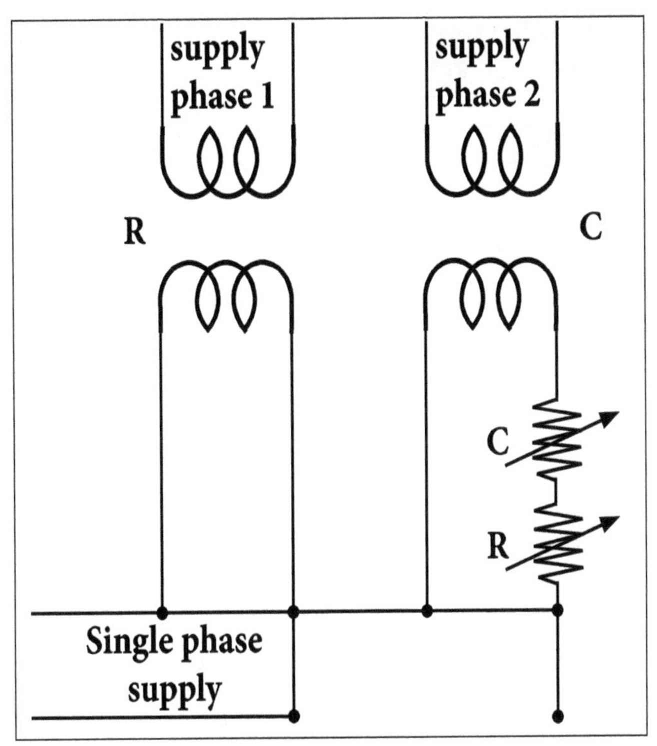

Phase splitting circuit.

The emf induced in the secondary winding of the mutual induction M assuming M to be free from eddy current effects will lag 90° in phase behind the current in the primary winding, i.e., in the quadrature potentiometer slide-wire. Also if 't' is the primary current, then the emf induced in the secondary is $2\pi \times$ frequency $\times$ M $\times$ i where M is the value of the mutual inductance. Therefore, for given values of frequency and mutual inductance, the induced emf when i has the standard value 50 mA, can easily be calculated.

Now,

$$\text{if } f = 50 \text{ Hz and } M = 0.0636 \text{ H the secondary induced.}$$
$$\text{emf} = 2\pi \times 50 \times 0.0636 \times 0.05 = 1.0 \text{ V, when i has the standard value.}$$

The slide-wire of the in-phase potentiometer is thus set to this calculated value of induced emf in the secondary of M (the slide-wire current being maintained at its standard value) and rheostat R and capacitor C are adjusted until exact balance is obtained.

For balancing the current in the quadrature potentiometer slide-wire must be both equal to the standard value and also must be exactly 90° out of phase with the current in the in-phase slide-wire. This latter condition follows from the fact that the emf in the secondary of M lags 90° in-phase behind the primary current, and, therefore, for this emf to be in-phase with the voltage drop across a portion of the in-phase slide-wire, the current in the primary of M must be in exact quadrature with the current in this in-phase slide-wire. Any difference in polarity between the two circuits is corrected by the sign-changing switches S_1 and S_2.

These adjustment having been made, the unknown voltage is switched in by means of the selector switch S_3. In this position of S_3, the two slide-wire circuits are in series with one another and with the vibration galvanometer. Balance is obtained by adjusting both pairs of sliding contacts (bb' and cc') together with the reversal of switches S_1 and S_2, if necessary.

At balance, the reading of the slide-wire of the in-phase potentiometer, together with the position of S_x, gives the magnitude and sign of the in-phase component of the unknown voltage, while the reading of the quadrature potentiometer, with the position of the switch gives the magnitude and sign of the quadrature component.

For example, if both S_1 and S_2 are in the positive position and V_1 and V_2 are the in-phase and quadrature components respectively of the unknown voltage V, then the phase of V is as shown in figure, while its magnitude is $\sqrt{V_1^2 + V_2^2}$.

Errors

The errors which may occur in using this potentiometer may be due to:

- Slight differences in the reading of the reflecting dynamometer instrument on AC as compared with the reading on DC. Such errors may cause the standard current value on AC to be slightly incorrect.
- An error in the nominal value of the mutual inductance M would cause the current in the quadrature slide-wire circuit to be somewhat different from the standard value.
- Inaccuracy of the method of measuring the frequency, which again would cause an error in the quadrature slide wire standard current value.
- The fact that inter-capacitance, earth capacitance and mutual inductance effects are present in the slide-wire coils and affect the potential gradient.
- Standardization of the potentiometer is based upon an rms value of current, while the potential balances on the slide-wires are dependent upon the fundamental only.

Quadrature Adjustments of Currents

The action of the phase-splitting circuit may be understood by referring to figure, which gives an equivalent circuit for the two potentiometers. R_1 and L_1 are the equivalent resistance and inductance of the in-phase potentiometer circuit and R_2 and those of the quadrature potentiometer. When the potentiometer currents are equal and in quadrature, $I_2 = jI_1$.

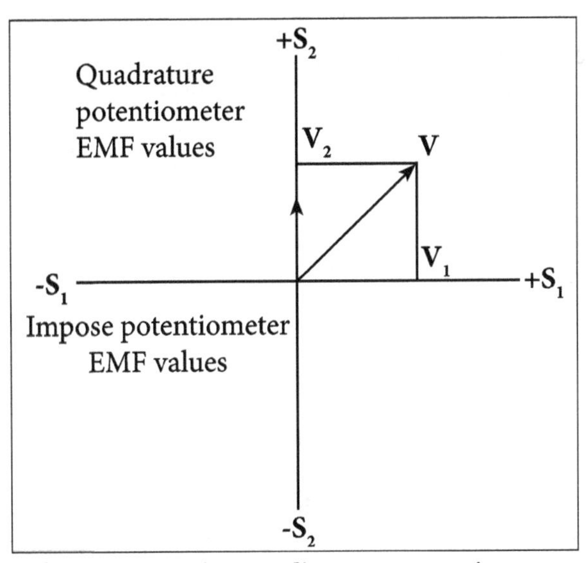

Phasor diagram for co-ordinate type potentiometer.

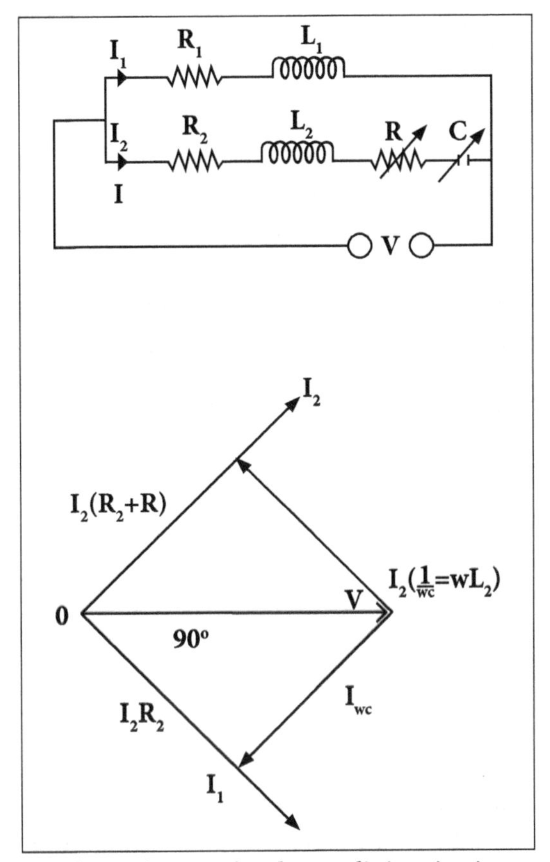

Phasor diagram for phase-splitting circuit.

or $$\frac{V}{R+R_2+j(\omega L_2-1/\omega C)}=j\frac{V}{R_1+j\omega L_2}$$

$$\text{Therefore, } R_1 + j\omega L_1 = j(R + R_2) - \left(\omega L_2 - \frac{1}{\omega C}\right)$$

By separating real and imaginary parts, we obtain the conditions for phase splitting as:

$$R_1 + \omega L_2 = \frac{1}{\omega C}$$

$$\text{and } \omega L_1 - R_2 = R$$

The phase splitting is adjusted by means of R and C.

This phase splitting has been used in later forms of the AC potentiometer in place of the original quadrature device described by DCG all, which uses a transformer and variable resistor.

Vernier Potentiometer

By using basic simple potentiometers, the precision of 100 μV for readings up to 1.6 V can be obtained. Using such instruments it is very difficult to get accurate readings mainly due to the non-uniformity of slide wire and maintaining good potential contacts. Some of the practical applications demand highly precised and accurate measurements. The limitations due to the slide wire in potentiometers are eliminated in the vernier potentiometers. The instrument with normal range of 1.6 V with 10 AV precision and lower range of 0.16 V with 100 μV precision is as shown in the Figure.

The main difference in the simple potentiometer and vernier potentiometer is that it uses three measuring dials. The slide wire is not used in this type of the potentiometer. The main dial i.e. the first dial measures up to 15 V on (X1) range in steps of 0.1 volts. The second dial reads up to 0.1 V in steps of 0.001 volts on (X0.001) range. It consists of 102 studs. The third dial again with 102 studs measures from - 0.00001 to + 0.001 volts on (X0.00001) range. This third dial provides true zero and negative setting. The resistances of the second dial shunt two of the coils of the main dial. The moving arm of the second dial camel two contacts which are placed two studs away from each other.

The vernier potentiometer reads voltages of 10 μV on X1 range while 1 μV on X0.1 range. To have voltage reading of 0.1μV one more range of X0.01 may be provided. But it is not possible to read such small voltages as stray thermal and contact potentials in potentiometer, galvanometer and measuring circuits are uncontrollable. These potentials are of the order of one to several microvolts. These potentials can be reduced by properly selecting metals for resistors, terminals and connecting leads. Sometimes thermal shields are also used to enhance process of reduction of potentials.

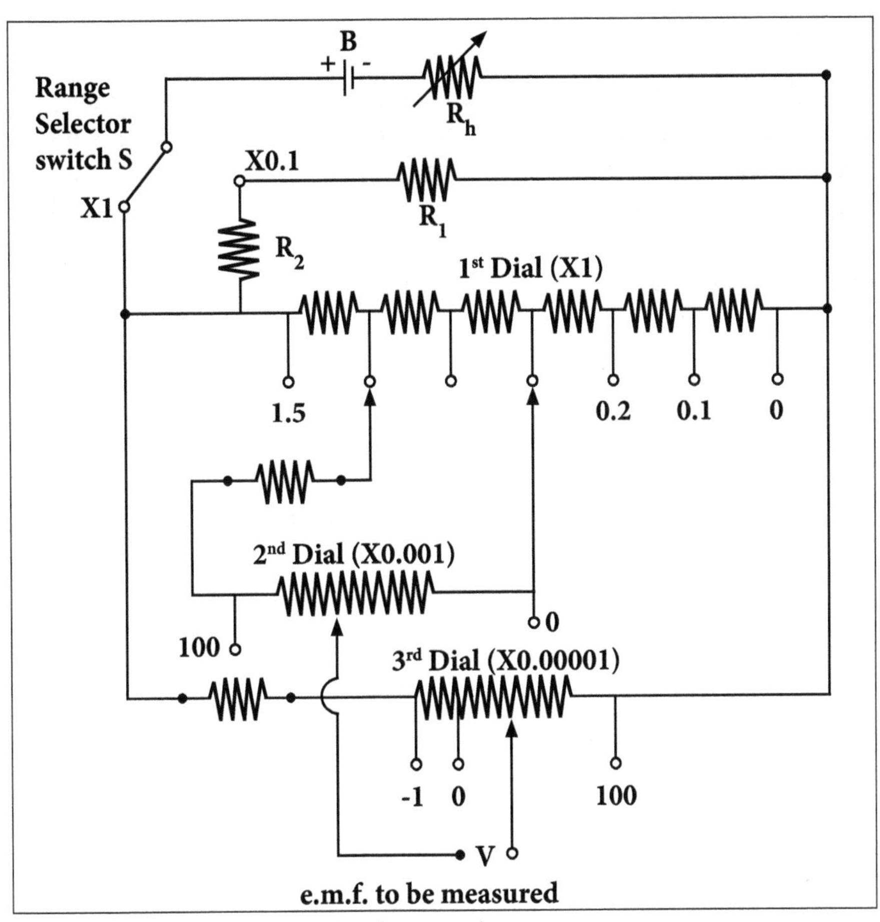

Vernier potentiometer.

Constant Potentiometer

A potentiometer in which the ratio of an unknown and a known potential are set equal to the ratio of two known constant resistances. Also known as Poggendorff's second method.

Brooks Deflection Potentiometer

Every potentiometers described so far are null instruments and a galvanometer serves as a null instrument to detect any departure from balance in the circuit. Thus, they are used where the circuit conditions are steady, i.e. the voltage to be measured is constant. Where circuit conditions are so unsteady that the voltage under measurement varies, it is often difficult or even impossible to obtain a balance. At the same time, for the measurement of moderate precision, exact balancing is time consuming and tedious. Brooks deflection potentiometer was developed to take care of these limitations. It is used for applications where the voltage to be measured is continuously changing. It is specifically used for calibration and testing of deflection instruments in the laboratories.

Construction

Figure shows the schematic diagram of a Brooks deflection potentiometer. It consists of only one or two main dials consisting of decade resistance boxes. A center zero type galvanometer is connected in the circuit to indicate deflection. The galvanometer is connected in series with compensating resistance R. A working battery E is connected in series with a rheostat R_h. The rheostat R_h is used to adjust the potentiometer current taken from the working battery E. AB represents the dials and slide-wire of the potentiometer while C and D are slide contacts.

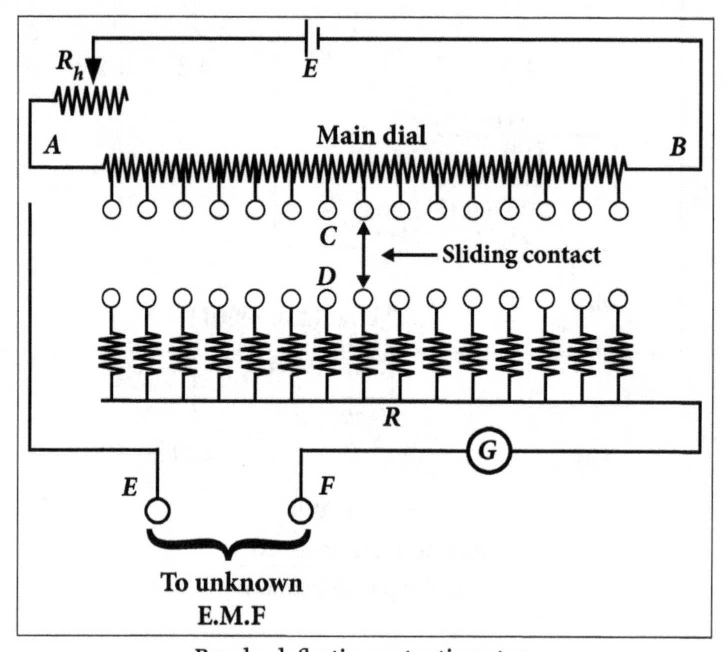

Brooks deflection potentiometer.

Working in Brooks deflection potentiometer, an approximate balance is obtained and the greater portion of the voltage is read from the setting of the slide contacts. The remaining varying portion is measured from the deflection of a galvanometer calibrated in millivolt. For the galvanometer to give a deflection proportional to the output of balance voltage, it is essential that the galvanometer circuit resistance is constant.

This is done by compensating resistance R, as shown in Figure. The values of the compensating resistance are so chosen that the resistance of the potentiometer circuit as viewed from terminals EF remains constant irrespective of the position of the sliding contacts. This means that current through the galvanometer will always be proportional to the out of balance current, whatever may be the setting of the main dials.

Thus, the galvanometer scale can be calibrated to read out of balance e.m.f. directly. The value of an unknown e.m.f. is obtained by adding the galvanometer reading to the main dial setting. The main dial setting is kept nearly equal to the e.m.f. being measured.

Drysdale-Tinsley Polar Type AC Potentiometer

Being a polar type, Drysdale - Tinsley a.c. potentiometer measures unknown e.m.f. In terms of its magnitude and phase angle. It consists of basic D.C. potentiometer along with some auxiliary components such as, Drysdale phase shifter and electrodynamometer type ammeter. Let us study Drysdale phase shifter construction first.

Drysdale phase shifter is also called phase shifting transformer. It consists of a ring shaped laminated steel stator. This sector is wound with either a two phase or three phase winding. Inside it there is a laminated rotor keeping some air gap between it and stator. The rotor consists of a winding provided in the slot which supplies voltage to slide wire circuit of potentiometer. The connection of Drysdale phase shifter with the circuit is as shown in the Figure.

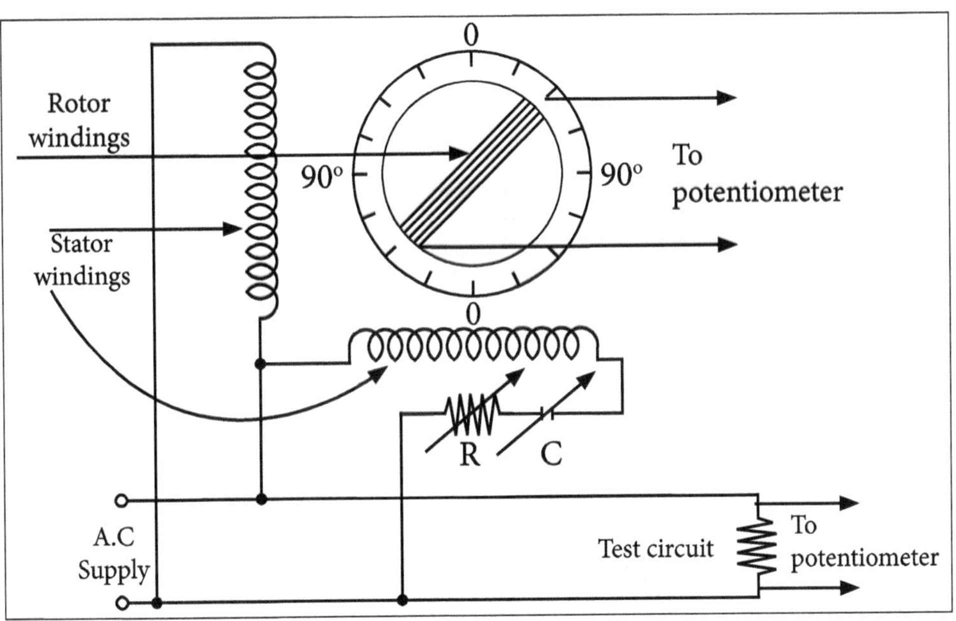

Drysdale phase shifter connection diagram.

When current flows through stator winding, a rotating field is produced inducing e.m.f. in the rotor. The phase of rotor current can be changed through any angle relative to stator supply voltage by rotating rotor. Thus the change in the phase of secondary e.m.f. is equal to the angle through which rotor is moved from its original zero position. It is very important to arrange windings such that even though the magnitude of e.m.f. induced in rotor is changed, phase remains unchanged.

Thus we can directly read the phase angle with the help of scale fixed on top of the instrument. The variable capacitor and resistor shown in the circuit diagram are so adjusted that exact quadrature component between the two stator winding currents is obtained. An electrodynamometer type ammeter is used to measure a.c. as well as D.C. currents during the standardization of an a.c. potentiometer.

2.4 Electro-Dynamometer, Induction Type Wattmeter and Power Factor

2.4.1 Electro-Dynamometer and Induction type Wattmeter

Single Phase Wattmeter:

The construction of electro-dynamometer wattmeter is similar to that of electrodynamometer ammeter and voltmeter. This type of wattmeter consists of a fixed coil which is connected in series with the load and it carries the current through the load.

Hence the fixed coil is also called as field coil or current coil, whereas the moving coil in the wattmeter is connected across the load and it carriers the current proportional to the voltage across the load. Hence the moving coil is also called as potential coil or pressure coil (or) voltage coil (or) P.C.

Construction Detail:

The constructional diagram of the electrodynamometer type wattmeter used for 1 φ power measurement consists of fixed coil.

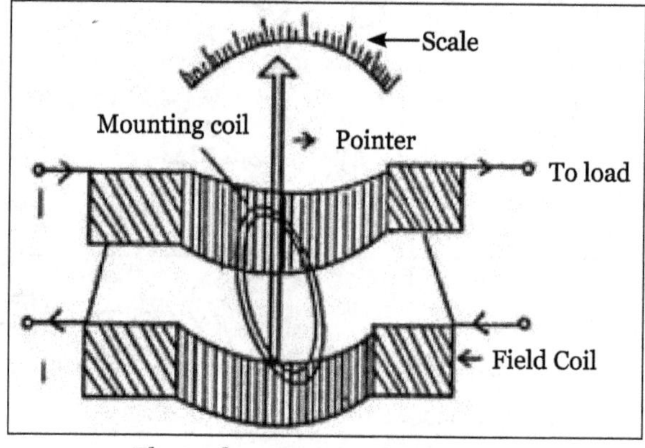

Electrodynamometer Wattmeter.

Fixed Coil:

Fixed coils are the field coils which carry the load current of the circuit. Generally they are divided into the halves but connected in series. The fixed coils are wound with heavy wire with less number of turns in order to have low resistance and hence low voltage drop across the meter.

The fixed coil also called as current coils and is always connected in series with the load and they are properly laminated to avoid eddy current losses in conductors when carrying heavy current, the maximum current range of wattmeter is 20A. However, for power measurements involving large load It is better to use a 5A wattmeter in conjunction with a current transformer of suitable range.

Moving Coil:

The moving coil is generally attached to the spindle which is connected to the pointer. The moving coil also called as pressure coil. The maximum load current flows through the load instead of flowing through the voltmeter connected across the load. A series resistor is used in the voltage circuit in order to limit the current to small Value in the order of 100 mA.

Both fixed and moving coil are air cored, the voltage rating of the wattmeter is limited to 600 V. However, for higher voltage range, a potential transformer can be used to step down the voltage.

Control Torque:

Control torque is provided by springs as it is a electrodynamometer type instrument.

Damping:

Air friction damping is used.

Pointer and Scale:

This type of instrument have mirror type scales and knife edge pointers to avoid parallax error while reading.

Torque Equation:

The figure shows the electric circuit of an electrodynamometer wattmeter.

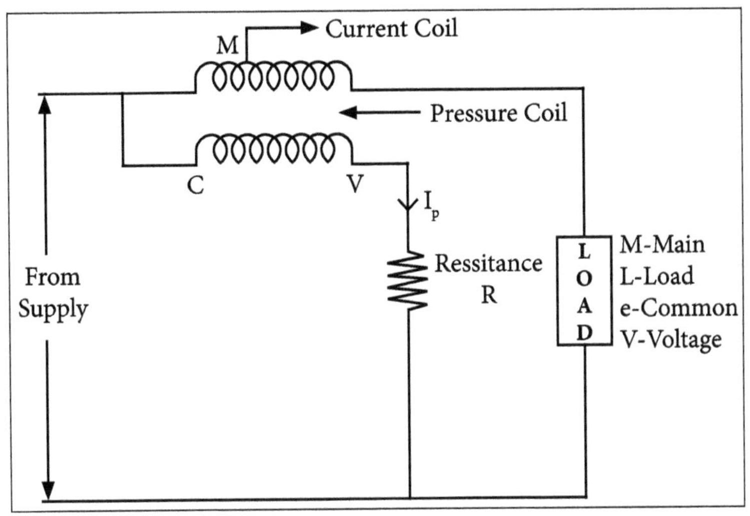

Electric circuit of an electrodynamometer wattmeter.

The instantaneous torque of an electrodynamometer instrument is given by:

$$T_i = i_1 i_2 \frac{dM}{d\theta} \qquad ...(1)$$

Where,

T_i = Instantaneous torque is N - m.

i_1 and i_2 = Instantaneous value of current in two coils in A.

Assumption: As the pressure coil has very high resistance, it is assumed to be purely resistive by neglecting the inductance.

W.K.T.

$$i\rho = \frac{V}{R} = \frac{\sqrt{2}V}{R} \sin \omega t = \sqrt{2}I \quad \sin \omega t \qquad ...(2)$$

As the current through the current coil (i.e.) lags the voltage across R_c by an angle φ:

$$i_C = \sqrt{2} I \sin(\omega t - \varphi) \qquad ...(3)$$

Let instantaneous torques is:

$$Ti = i_\rho i_c \cdot \frac{dM}{d\theta} \qquad ...(4)$$

Substituting the equation (2) and (3) in (4), we get:

$$T_i = \sqrt{2}I_P \sin \omega t \times \sqrt{2} \, I \sin(\omega t - \phi) \frac{dH}{d\theta} \qquad ...(5)$$

On simplification, we get:

$$T_i = I_P I \left[\cos\phi - \cos\left(2\omega t - \phi\right)\right] \frac{dH}{d\theta} \qquad ...(6)$$

Average deflecting torque is:

$$T_d = \frac{1}{T}\int_0^T T_i \, d(\omega t)$$

$$T_d = \frac{1}{T}\int_0^T l_p l \left[\cos\phi - \cos[2\omega t - \phi] \frac{dM}{d\theta} \cdot d(\omega t)\right]$$

On simplification:

$$T_d = I_p \cos\phi \frac{dM}{d\theta}$$

$$T_d = \frac{V_I}{R_P} \cos\theta \frac{dM}{d\theta} \left[\therefore I_P = \frac{V}{R_P}\right]$$

Controlling torque exerted by the spacing is given by, $T_C = K_s \theta$.

At balance position, $T_C = T_d$.

Hence:

$$K_S\,\theta=\frac{VI}{R_P}\cos\phi\frac{dM}{d\theta}$$

$$\theta=\frac{VI}{R_P\,K_S}\cos\phi\frac{dM}{d\theta}$$

$$\theta\,\alpha\,VI\cos\varphi \quad ...(7)$$

From equation (7) it is clear that the deflection of the pointer is the direct indication of single phase power.

Single Phase Induction Type Energy Meter

Basic Principle

It is an integrating type instrument which works on the principle of induction. The alternating fluxes induces the generation of eddy current in the moving system which interacts with each other and produces driving torque which causes the aluminum disc to rotate and thus records the energy.

Constructional Details

Four main parts of the induction type energy meter are:

(i) Driving System

- The driving system of the energy meter consists of two electromagnets, whose core is made up of silicon steel laminations. The coil of the electromagnet excited by the load current is called current coil and the corresponding electromagnet is called series magnet.
- The coil of the second electromagnet which is connected across the supply carries a current proportional to the supply voltage and is called pressure coil and the corresponding electromagnet is called shunt magnet.
- The function of the adjustable copper shading in shunt magnet is to bring the flux produced by the shunt magnet exactly in quadrature with the applied voltage.

(ii) Moving System

- The moving system consists of an aluminium disc mounted on a light alloy shaft. The disc is positioned in the air gap between the series and shunt magnets.
- The moving system is connected to a hardened steel pivot, is supported by a jewel bearing.
- In this type of energy meter as there is no controlling torque, continuous rotating of the disc is produced due to the driving torque only.

(iii) Braking System

- The braking system consists of a permanent magnet positioned near the edge of the aluminium disc.
- The aluminium disc moves in the field of this magnet and thus provides a braking torque.
- By adjusting the position of the permanent magnet, braking torque can be adjusted.

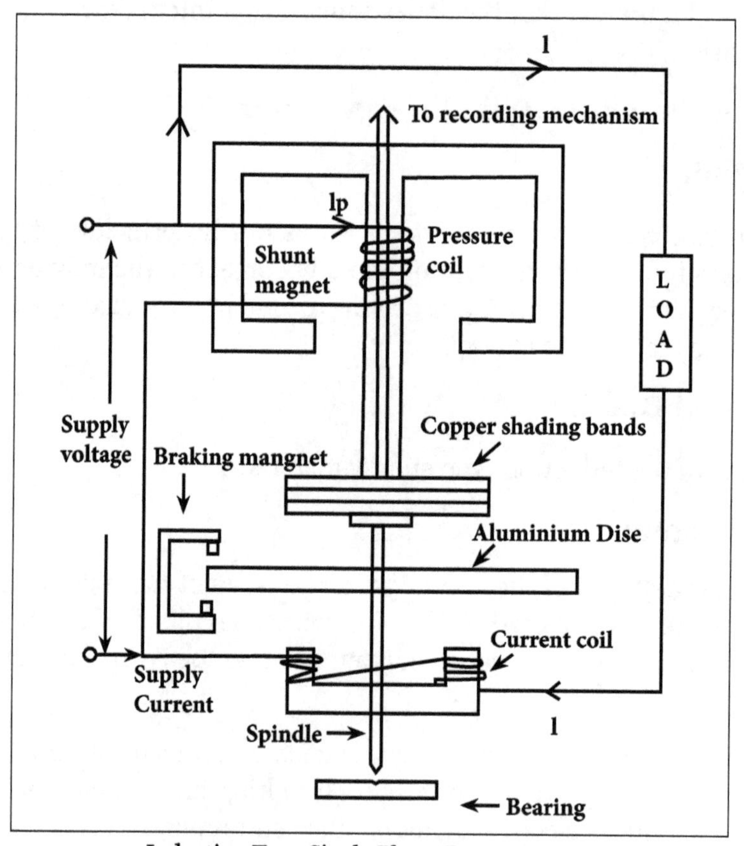

Induction Type Single Phase Energy Meter.

(iv) Registering/Counting Mechanism

- The function of a registering or round mechanism is to record continuously a number which is proportional to the revolutions made by the moving system.
- The pointer rotates on round dials which are marked with ten equal divisions.

Working:

- When the pressure coil wound on the shunt magnet is connected across the supply voltage, it carries a current I_p proportional to the supply voltage, thus producing an alternating flux φ_p.

- Flux φ_p induces a emf E_{ep} in the disc, produces eddy current I_{ep}. The current on the series magnet carries load current I, it produces an alternating flux φ_X, it induces an emf E_{ec} in the disc, produces an eddy current I_{ec}.
- I_{er} interacts with φ_X produces a torque T_1. Similarly I_{ec} interacts with φ_p produces T_2, the net torque is the difference of the above two torques.

Let,

V – Supply voltage,

I_p – Current through the pressure coil which is proportional to supply voltage,

I – Load current,

ϕ – Phase angle of load,

Φ_p –Flux produced by current through pressure coil,

Φ_X –Flux produced by current through current coil,

E_{ep} –Eddy emf induced by flux Φ_p,

E_{ec} –Eddy emf induced by flux Φ_X,

I_{ep} – Eddy current due to flux Φ_p,

I_{ec} – Eddy current due to flux Φ_X.

Net driving torque, $T_d \propto T_2 - T_1$.

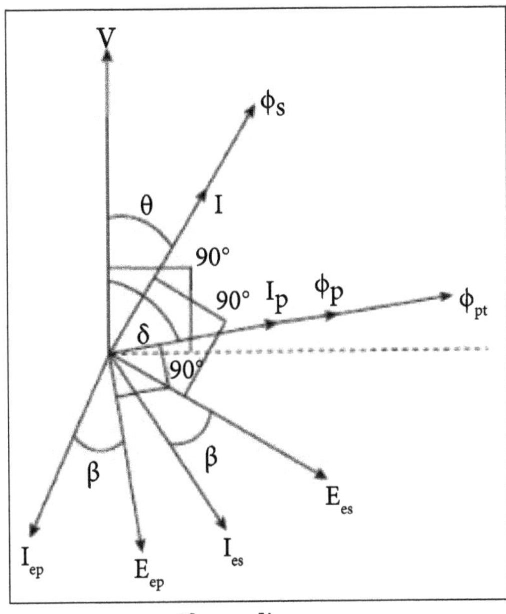

Phasor diagram.

2.4.2 Single Phase and Polyphase Induction Type Watt-Hour Meters

Induction Type Wattmeter

These types of watt-meters operate on same working principle on which the induction type ammeter and voltmeter operates. These instruments can be used only with ac supply while dynamo-meter type watt meters can be used on either ac or dc supply system. Induction type watt-meters are useful only when supply and frequency remains constant. Since both the coils i.e. current coil and the pressure coils are necessary in such instrument, it is not essential to use shaded pole principle. Because for producing deflecting torque, two fluxes are essential with suitable phase angle and it would be available from these two coils.

Construction

A watt-meter has two laminated electromagnet, one of which is excited by load current or definite fraction of it, and is connected in series with the circuit, known as series magnet and the other is excited by the current proportional to applied voltage or fraction of it and is always connected across the supply, known as shunt magnet. An aluminum disc is mounted so that it cuts the fluxes produced by both the magnets. As a result of which, two e.m.fs are produced which induces two eddy currents in the disc. C - Magnet is used to provide necessary damping torque to the pointer, to damp out oscillations. Deflecting torque is produced due to interaction of these eddy currents and inducing flux. Copper shading bands are provided either on central limb or on the outer limb of the shunt magnet, and can be so adjusted as to make the resultant flux in the shunt magnet lag behind the applied voltage by 90° Both the watt-meters are provided with spiral springs A and B, for producing controlling torque to counter balance the deflecting torque. In Fig. 13.2 the spiral spring and damping magnet is omitted for simplicity. The scale of such type instruments is quite uniform and extends over an angle of 3000 Currents up to 100° A can be handled by these watt-meters directly where as beyond this current transformers are used. Two types of induction type watt meters are available. Line diagrams of both of the types are detailed in Figure a and b.

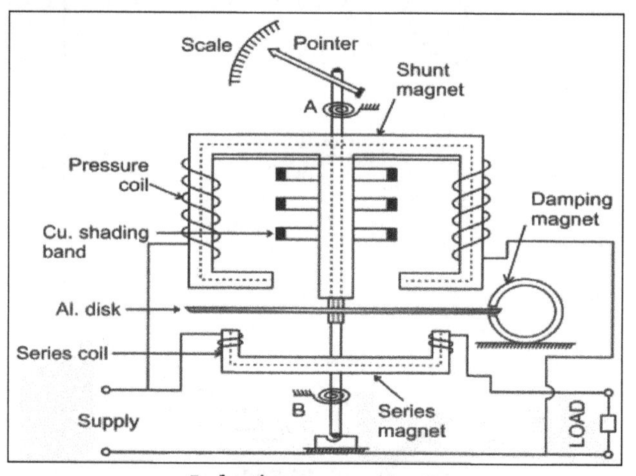

a. Induction type wattmeter.

In the form of the instrument shown in Figure a, two pressure coils are connected in series in such a way that both of them send flux through the central limb. The series magnet also carries two small current coils connected in series and wound so that they magnetized their respective cores in the same direction. Correct phase displacement between the fluxes produced by series and shunt magnet is obtained by the adjustment of copper shading band on the central limb.

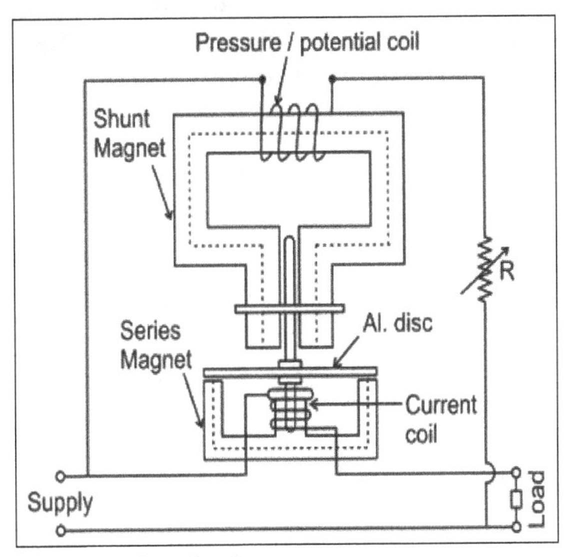

b. Induction type wattmeter.

In Figure b, there is only one pressure and one current coil. Two projecting poles of shunt magnet are surrounded by a copper shading band whose position can be adjusted for correcting the phase of the flux of this magnet with the applied voltage. The pressure coil circuit of induction type instrument is made as inductive as possible so that the flux of the shunt magnet may lag nearly by 90o behind the applied voltage.

Advantages:

The advantages of induction watt meters are the same as those of induction ammeters o long scale, freedom from effects of stray field, and have effective damping torque.

Disadvantages:

Following are the disadvantage of the induction type instruments:

- Change in temperature causes variation in resistance of the moving element, affects the eddy currents therein, and so the operating torque. The error due to this in part offset by a balancing effect due to change in temperature of the windings.
- Change in frequency from that of calibration value causes variations in both the reactance of the voltage coil circuit, which is highly inductive, and also in amount of compensation from the phase compensating circuit. Within the limits of frequency variation met within practice on mains, this last error in not important.

Induction Type Single Phase Watt Hour Meter

A watt hour meter is used to sum up the total energy consumed by a consumer during a period so that it can be charged for actual energy consumed. The working principle, theory and advantage / disadvantages are most similar to single phase watt meter. The construction of single phase watt hour meter is also almost similar to single phase induction type watt meter as discussed. The pointer and spiral springs are replaced by wheel-train mechanism for summing of total energy consumed where as the damping magnet is replaced by braking magnet. The construction of this watt hour meter is shown in Figure c.

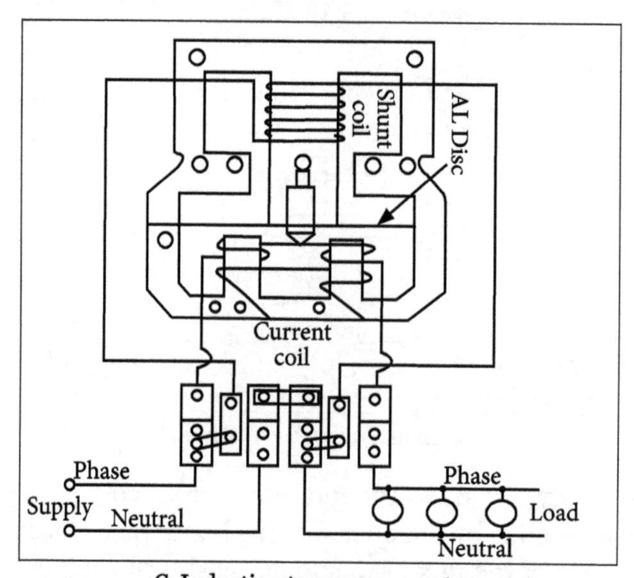

C. Induction type energy meter.

The brake magnet and recording wheel-train being omitted for the clear understanding of the diagram. The description of registering mechanism and braking system is detailed below.

Registering or Counting System

The registering or counting system essentially consists of gear train, driven either by a worm or pinion gear on the disc shaft, which turns pointers that indicate on dials the number of times disc has turned. The energy meter thus determines and adds together or integrates all the instantaneous power values so that the total energy used over a period is thus known. Therefore, this type of meter is also called an integrating meter.

Braking System

Braking of the disk is provided by a small permanent magnet, located diametrically opposite to alternating current magnets. The disk moves between the magnets gaps. The movement of rotating disc through the magnetic field crossing air gap sets up eddy currents in the disc that reacts with the magnetic field and exerts braking torque. By changing the position of the brake magnet or diverting some of the flux therefore, the speed of rotating disc can be controlled. Creep error can be rectified by drilling a small hole in aluminum disc passing through the magnetic flux of braking magnet.

Poly-Phase Induction Watt-Hour Meters

The most common type of poly-phase induction watt-hour meter consists of two or more single-phase meters combined with the disc mounted on a common shaft and arranged to record on a single register. The performance of poly-phase meters is comparable to that of single-phase meters. Owing to the fact that two or more elements comprising the poly-phase meter are mechanically coupled by acting upon a common moving element, full-load adjustment for individual elements loses its significance.

The adjustments must include some additional features by which the torque of the individual element may be brought to equality. This equalizing adjustment is called balancing, which is sometimes secured by changing the air gap between voltage and current electromagnets of either element. Decreasing the air gap increases the torque contributed by the clement being adjusted. In other types of adjustments, balance is secured by moving an open loop of magnetic material in or out of the potential electromagnet air gap; or by a tilting movement of both potential and current electromagnets to alter their position relative to the discs.

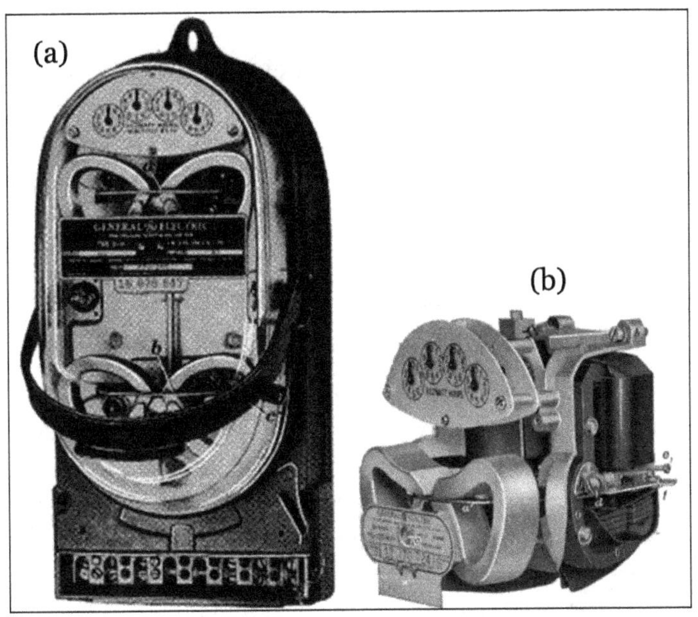

(a) A General Electric Company Type Two—Element Poly Phase Meter
(b) The Motive Elements of the Meters.

2.4.3 Frequency Meters

An instrument for measuring the frequency of periodic processes or oscillations. The frequency of mechanical vibrations is usually measured by means of mechanical vibration frequency meters and by the electrical meters equipped with transducers to convert the mechanical vibrations into electrical oscillations.

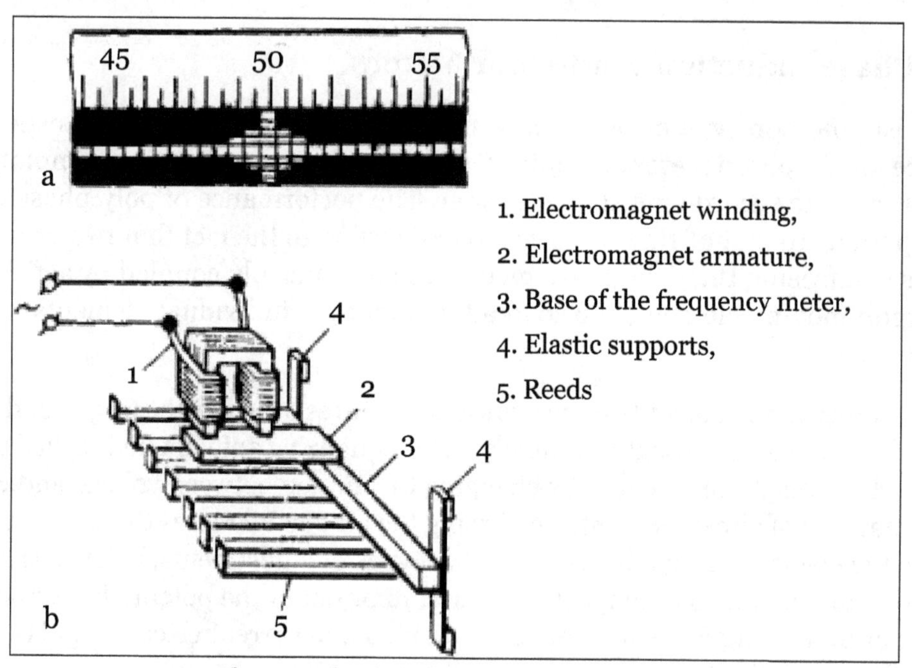

Electromechanical vibration frequency meter:
(a) scale, registering a reading of 50 Hz, (b) diagram of the instrument.

The simplest mechanical vibration frequency meter operates on the resonance principle and consists of a series of flexible reeds fastened at one end to the common base. The lengths and masses of the reeds are chosen in such a way that their natural vibration frequencies form specified discrete scale, from which the value of the frequency being measured is determined. When the mechanical vibrations act on the base of the meter, they cause the flexible reeds to vibrate. The highest vibration amplitude is observed on the reed whose natural vibration frequency is equal or close to the value of the frequency being measured.

The frequency of electrical oscillations is measured by means of electromechanical, electrodynamic, electronic, moving-iron, and moving-coil frequency meters. The simplest electromechanical type consists of an electromagnet and a series of flexible reeds as in the mechanical frequency meter on a common base that is attached to the armature of the electromagnet (Figure a). The electrical oscillations being measured are fed to the winding of electromagnet; the armature vibrations thereby produced are transmitted to the reeds, and the value of the frequency being measured is determined from the vibrations.

The principal element in the electrodynamic frequency meters is a ratio meter with an oscillatory circuit in one of its branches that is permanently tuned to average frequency for the measurement range of the given instrument (Figure b). When connected to the AC circuit, the moving part of the ratio meter is deflected by an angle proportional to the phase shift between the currents in windings of the ratio meter, which depends on the ratio of the frequency being measured to resonance frequency of the oscillatory circuit. The measurement error of electrodynamic frequency meters ranges from 10^{-1} to 5×10^{-2}.

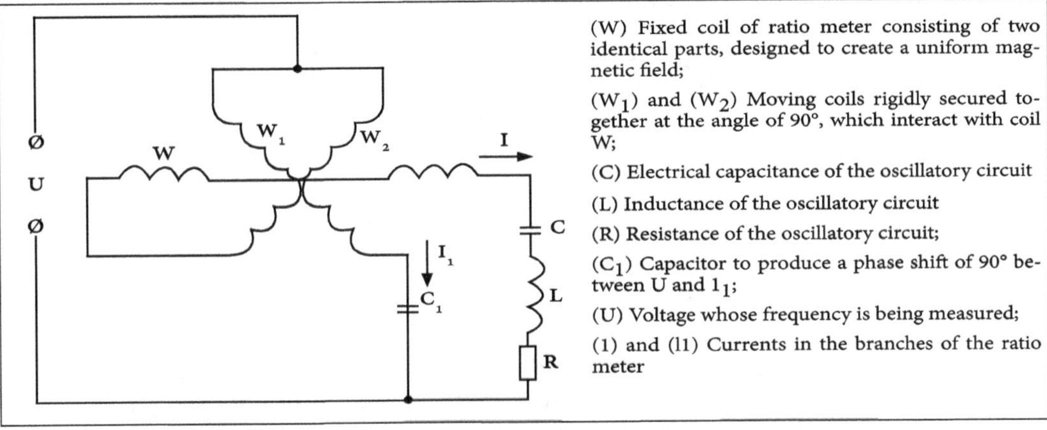

Schematic diagram of an electrodynamics frequency meter.

The frequency of electromagnetic oscillations in radio-frequency and microwave-frequency ranges is measured by the means of the electronic frequency meters (wave meters), such as the resonant, heterodyne, and digital types.

The operation of resonant-type frequency meter is based on comparison of the frequency being measured with the frequency of natural oscillations in a electrical circuit (or a microwave resonator) that is tuned to resonance with frequency being measured. The meter consists of a oscillatory circuit with a coupling loop that picks up the electromagnetic oscillations (radio waves), a detector, a amplifier, and a resonance indicator (Figure 3). During measurement, the circuit is tuned by means of the calibrated capacitor (or the plunger of a microwave resonator) to the frequency of electromagnetic oscillations being picked up until the resonance is achieved, as shown by the greatest deflection of the pointer on the indicator. The measurement error ranges between 5×10^{-3} and 5×10^{-4}.

In heterodyne frequency meters, the frequency being measured is compared with known frequency (or one of its harmonics) produced by an oscillator, or heterodyne. As the heterodyne frequency is tuned to frequency of the oscillations being measured, beats occur at output of a mixer in which frequencies are compared. After amplification the beats are indicated by the pointer on a instrument, by an earphone, or sometimes by the oscilloscope. The relative error of heterodyne frequency meters ranges from 5×10^{-4} to 5×10^{-6}.

Digital frequency meters or frequency counters are now widely used. Their operation involves a count of number of periods in the oscillations being measured during specified time interval. Frequency counters consist of a device that converts sinusoidal voltage of frequency being measured into a train of unidirectional pulses, a gate for pulses that opens for a certain time interval (usually from 10^{-4} to 10 sec), an electronic counter that registers number of pulses at the gate output, and digital display. Modern digital frequency counters operate over frequency range from 10^{-4} to 109 hertz with a relative measurement error from 10^{-9} to 10^{-11} and sensitivity of 10^{-2} volt. Such devices are used primarily for testing the radio equipment and, with various measuring transducers, for measuring the temperature, vibrations, pressure, strain, and the other physical quantities.

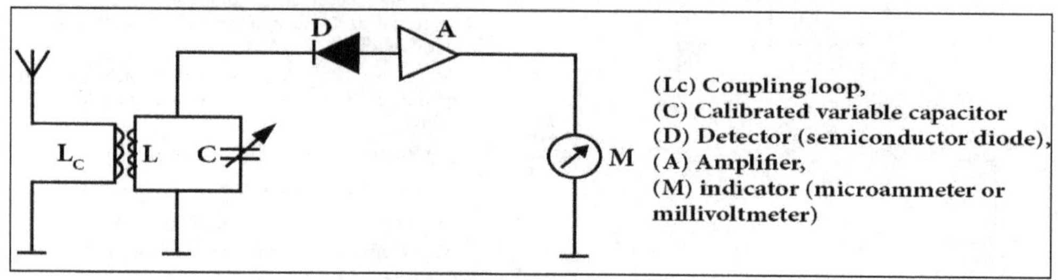

Schematic diagram of a resonant-type frequency meter.

Primary and the secondary frequency standards, which have an error in the range from 10^{-12} to 5×10^{-14}, function as a type of high accuracy reference frequency meters. The rotational speed of shafts in machines and the mechanisms is measured with a tachometer.

2.4.4 Power Factor Meters

Electrical Power Factor

In general power is the capacity to do work. In electrical domain, electrical power is the amount of electrical energy which can be transferred to some other form (heat, light etc) per unit time. Mathematically it is product of voltage drop across the element and current flowing through it. Considering first the DC circuits, having only DC voltage sources, the inductors and capacitors behave as short circuit and open circuit respectively in the steady state. Hence entire circuit behaves as resistive circuit and the entire electrical power is dissipated in the form of heat. Here the voltage and current are in same phase and the total electrical power is given by Electrical power = Voltage across element X Current through the element. Its unit is Watt = Joule/sec.

Now coming to AC circuit, here both inductor and capacitor offer certain amount of impedance given by:

$$X_L = 2\pi f L \ \text{ and } X_C = \frac{1}{2\pi f C}$$

The inductor stores electrical energy in the form of magnetic energy and capacitor stores electrical energy in the form of an electrostatic energy. Neither of them dissipates it. Further there is a phase shift in between voltage and current. Hence when we consider entire circuit consisting of resistor, inductor and capacitor, there exists some phase difference between the source voltage and current. The cosine of the phase difference is called electrical power factor.

This factor $(-1 < \cos\phi < 1)$ represents fraction of total power that is used to do the useful work. The other fraction of electrical power is stored in the form of magnetic energy or the electrostatic energy in inductor and capacitor respectively. The total power in this case is, Total electrical power = Voltage across element X current through the element. This is called apparent power and its unit is VA (Volt Amp) and denoted by 'S'. A fraction of this total electrical power which actually does our useful work is called as active power.

It is denoted as 'P'.

P = Active power = Total electrical power. $\cos\varphi$ and its unit is watt.

The other fraction of power is called reactive power.

This does no useful work, but it is required for the active work to be done.

It is denoted by 'Q' and is mathematically given by, Q = Reactive power = Total electrical power. $\sin\varphi$ and its unit is VAR (Volt Amp Reactive). This reactive power oscillates between source and load.

To help understand this all these power are represented in form of triangle:

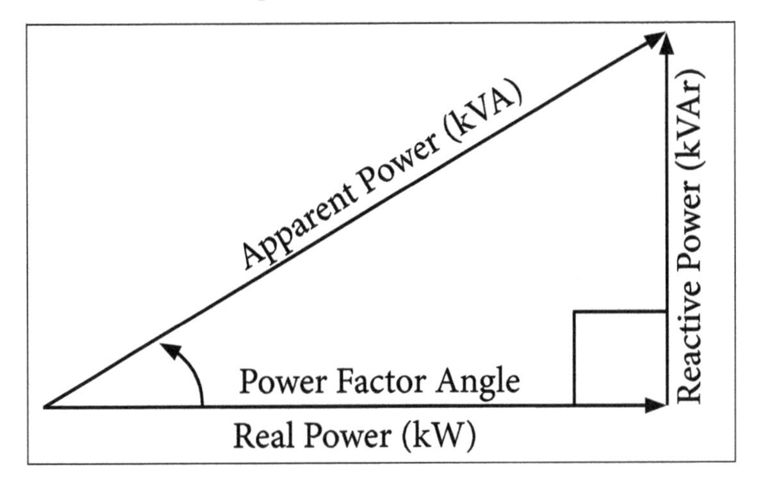

Mathematically, $S^2 = P^2 + Q^2$ and electrical power factor is given by active power / apparent power.

Power Factor Improvement

The term power factor comes into picture in AC circuits only. Mathematically it is the cosine of the phase difference between source voltage and current. It refers to the fraction of the total power (apparent power) which is utilized to do the useful work called active power.

$$\cos\phi = \frac{\text{Active power}}{\text{Apparent power}}$$

Need for Power Factor Improvement

The real power is given by the expression $P = VI\cos\varphi$. To transfer a given amount of power at certain voltage, electrical current is inversely proportional to power factor $\cos\varphi$. Hence higher the p f lower will be the current flow. A small current flow requires less amount of the cross sectional area of conductor and thus it saves the conductor and money.

From above relation we saw having poor power factor increases the current flowing in the conductor and thus copper loss increases. Further large voltage drop occurs in alternator, electrical transformer and the transmission & distribution lines which gives very poor voltage regulation.

Further the KVA rating of machines is reduced by having higher power factor as:

$$KVA = \frac{KW}{\cos\phi}$$

Hence, the size and cost of machine also reduced. So, electrical power factor must be maintained close to unity.

Methods of Power Factor Improvement

Capacitors: Improving power factor means reducing phase difference between voltage and current. Since majority of the loads are of inductive nature, they require some reactive power for them to function. This reactive power is provided by the capacitor or bank of capacitors installed parallel to load. They act as a source of the local reactive power and thus less reactive power flows through the line. Basically it reduces the phase difference between the voltage and current.

Synchronous Condenser: It is a 3 phase synchronous motor with no load attached to its shaft. The synchronous motor has the characteristics of operating under any power factor which is leading, lagging or unity depending upon excitation. For inductive loads, synchronous condenser is connected towards the load side and is overexcited. This makes it behave like capacitor. It draws the lagging current from the supply or it supplies the reactive power.

Phase Advancer: It is an ac exciter mainly used to improve pf of induction motor. They are mounted on shaft of motor and is connected in rotor circuit of the motor. It improves the power factor by providing exciting ampere turns to produce required flux at slip frequency. Further if ampere turns are increased, it can be made to operate at the leading power factor.

Power Factor Calculation

In power factor calculation, we measure source voltage and current drawn using a voltmeter and ammeter respectively. A wattmeter is used to get the active power. Now, we know that $P = VI\cos\varphi$ watt.

$$\text{From this } \cos\phi = \frac{P}{VI} \text{ or } \frac{\text{Wattmeter reading}}{\text{Voltmeter reading} \times \text{Ammeter reading}}$$

Hence, we can get the electrical power factor. Now we can calculate the reactive power $Q = VI\sin\varphi$ VAR. This reactive power can now be supplied from the capacitor installed in parallel with load in local. Value of the capacitor is calculated as per following formula:

$$Q = \frac{V^2}{X_C} \Rightarrow C = \frac{Q}{2\pi f V^2} \text{ farad}$$

3

Transformers, Electronic Instruments, Oscilloscope, Counters and Analyzers

3.1 Current Transformer and Potential Transformer

Current Transformer (C.T.)

The large alternating currents that cannot be sensed or passed through normal ammeters and current coils of wattmeters, energy meters can easily be measured by use of current transformers along with normal low range instruments.

A transformer is a device that consists of two windings called primary and secondary. It transfers energy from one side to another with suitable change in the level of current or voltage. A current transformer basically has a primary coil of one or more turns of heavy cross-sectional area. In some, the bar carrying high current may act as a primary. This is connected in series with line carrying high current. This is shown in the Figure (a). The bar type primary is shown in the Figure (b). The secondary of the transformer is made up of a large number of turns of fine wire i.e. having a small cross-sectional area. This is usually rated for 5 A current. This is connected to the coil of normal low range meter.

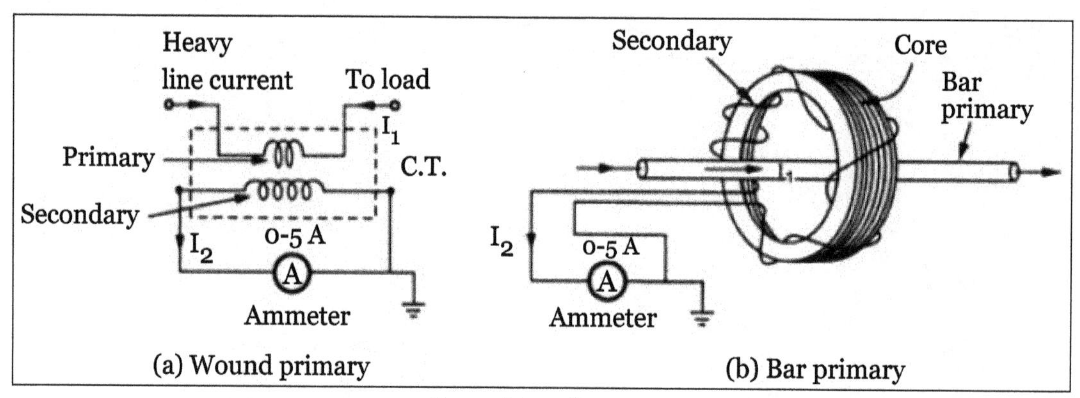

Current transformer.

These transformers are basically step up transformers that is stepping up a voltage from primary to secondary. Hence obviously current considerably gets stepped down from primary to secondary. For example C.T. is of 500: 5 range i.e., if primary current is 500 A it will be reduced to 5 A on secondary. But it steps up the primary voltage 100 times.

$$\frac{I_1}{I_2} = \frac{N_1}{N_2}$$

This is the current and number of turns relationship for the current transformers. Hence if current ratio of C.T. is known and meter reading is known, the actual high line current value can be determined.

It is very important that the secondary of C.T. should not be kept open. Either it should be shorted or it must be connected in series with a low resistance coil such as current coils of the wattmeter, coil of ammeter etc. If it is left open, then current through secondary becomes zero hence the ampere turns produced by the secondary which generally oppose primary ampere turns becomes zero. As there is no counter m.m.f., unopposed primary m.m.f. (ampere turns) produces high flux in the core. This produce excessive core losses, heating the core beyond limits. Similarly heavy emfs will be induced on the primary and secondary side. This may damage the insulation of the winding. This is dangerous from operator point of view as well. It is usual to ground the C.T. on the secondary side to avoid a danger of shock to the operator.

Potential Transformer (P.T.)

The basic principle of these transformers is same as the current transformers. The high alternating voltages are reduced in a fixed proportion for the measurement purpose with the help of potential transformers. The construction of these transformers is similar to the normal transformer. These are extremely accurate ratio step down transformers. The windings are low power rating windings. Primary winding consists of large number of turns while the secondary has less number of turns and usually rated for 110 V, irrespective of the primary voltage rating. The primary is connected across the high voltage line while secondary is connected to low range voltmeter coil. One end of the secondary is always grounded for safety purpose. The connections are shown in the Figure.

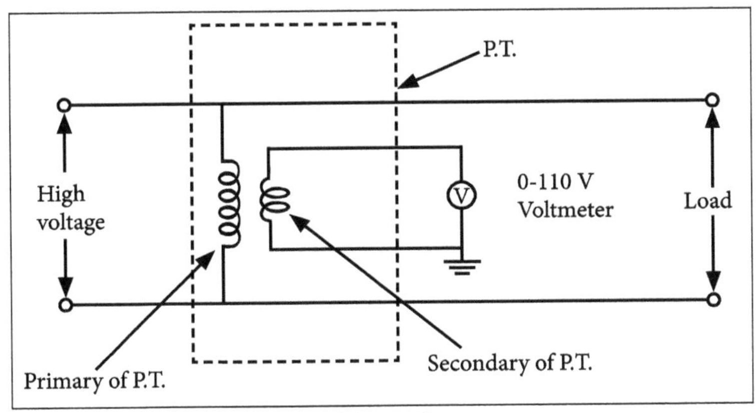

Potential transformer.

As a normal transformer, its ratio can be specified as:

$$\frac{V_1}{V_2}=\frac{N_1}{N_2}$$

So if voltage ratio of P.T. is known and the voltmeter reading is known then the high voltage to be measured can be determined.

Ratio Error

In practice it is said that current transformation ratio I_2 / I_1 is equal to the turns ratio N_1 / N_2. But actually it is not so. The current ratio is not equal to the turns ratio because of magnetizing and core loss components of the exciting current. It also gets affected due to the secondary current and its power factor. The load current is not a constant fraction of the primary current. Similarly in case of potential transformers, the voltage ratio V_2 / V_1 is also not exactly equal to N_2 / N_1 due to the factors mentioned above. Thus the transformation ratio is not constant but depends on the load current, power factor of load and exciting current of the transformer. Due to this fact, large error is introduced in the measurements done by the instrument transformers. Such an error is called ratio error.

The ratio error is defined as:

$$\%\ \text{Ratio error}=\frac{\text{Nominal ratio}-\text{Actual ratio}}{\text{Actual ratio}}\times 100$$

$$\%\ \text{Ratio error}=\frac{K_n-R}{R}\times 100$$

Phase Angle Error

In the power measurements, it is must that the phase of secondary current is to be displaced by exactly 180° from that of the primary current for C.T. While the phase of secondary voltage is to be displaced by exactly 180° from that of primary voltage, for P.T. But actually it is not so. The error introduced due to this is called phase angle error. It is denoted by angle θ by which the phase difference between primary and secondary is different from 180°.

The phase angle error is given by:

$$\theta=\frac{180}{\pi}\left[\frac{I_m\cos\delta-I_C\sin\delta}{n I_S}\right]\text{degrees}$$

Approximate results: In practice, the loads are inductive and δ is positive and very small.

$$\sin\delta=0 \text{ and } \cos\delta=1$$

$$R = n + \frac{I_C}{I_S}$$

$$\theta = \left(\frac{180}{\pi}\right)\left(\frac{I_m}{n I_S}\right) \text{ degrees}$$

Interns of I_p, these can be written using $n = \frac{I_P}{I_S}$.

$$R = n + \frac{n I_C}{I_p}$$

$$\theta = \left(\frac{180}{\pi}\right)\left(\frac{I_m}{I_p}\right) \text{ degrees}$$

Current Transformer Design

An effective design of a Ring Type C.T. may be produced first time using the following procedure, without any previous experience. The C.T. will induce current in its secondary winding which serves to completely oppose the magnetizing effect of the primary current, except for that small proportion required to magnetize the core. This core magnetizing component will then be the only source of error if the secondary current is to be used as a measure of the primary current. Making two assumptions i.e. that the CT has no leakage reactance and that its burden is purely resistive, the vector diagram for a one-to one ratio CT will look like this:

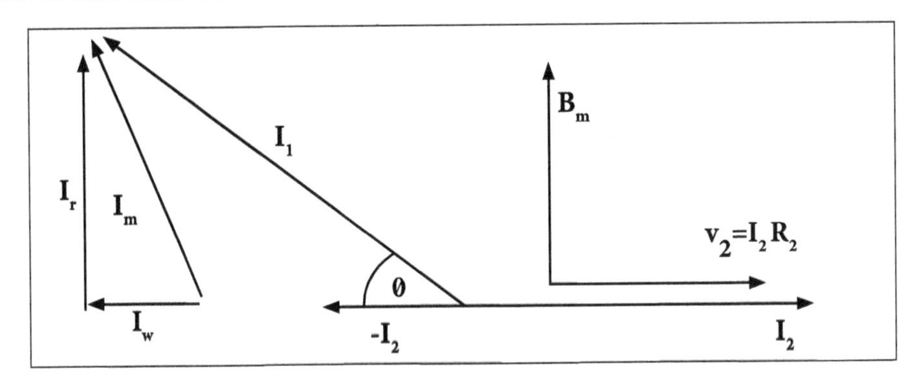

N_2 = No. of secondary turns.

V_2 =Secondary Voltage.

R_b = Burden Resistance.

I_1 = Primary Current.

I_2 = Secondary Current.

I_m = Excitation current.

I_r = Reactive component of I_m.

I_w = Watt loss of component I_m.

e = Ratio Error.

From this diagram, the primary current I1 differs from the secondary I2 in magnitude and phase angle. The angle error θ is Sin -1 Ir/I1 and the magnitude of,

$$I_1 = \left[(I_2 N_2 + I_w)^2 + I_r^2 \right]^{1/2}$$

In practice, the angle θ is so small as to allow the approximations $I_1 N_2$ + Iw and θ = I_r / I_1 radians, i.e. the current error is due to the watt loss component of the excitation current and the phase error is proportional to the reactive component I_r. The ratio error can be corrected by an amendment to the turns ratio, the secondary winding being reduced by several turns or fractions of a turn. Because of the non - linearity of excitation characteristics, such corrections do not maintain accuracy as the current changes, and a choice must be made that gives good balance over the whole range of current. Cores can be supplied with drilled holes, enabling fractions of a turn to be wound. The phase angle error, on the other hand, cannot be corrected, being a function of reactive component of the excitation characteristics which vary widely over the current range and must take priority in design of the transformer and choice of core.

The procedure is best described by considering an example, as follows:

Transformer Specification

Ratio 150/1

50Hz. Burden 2.5Va at Power Factor =1.0

Accuracy BS.3938, Class 0.5

Insulation Level – 11 Kv

Maximum Permissable Error:

From 10% to 20% of rated current	Ratio error Phase displacement	1% to 60 minutes
From 20% to 100% of rated current	Ratio error Phase displacement	0.75% 45 minutes
From 100% to 120% of rated current	Ratio error Phase displacement	0.5% 30 minutes

Internal Diameter

The I.D. of the core is fixed by physical consideration of the primary conductor and insulation, plus allowance for the secondary winding and core insulation. The main insulation is invariably placed on the primary conductor so that a 20 mm dia. conductor insulated for 11 Kv will have an overall diameter of about 40mm. The Secondary winding and core insulation for a nominal 660 volts lead to the choice of the core I.D. of 60mm. Assuming a maximum O.D. of 110mm, the mean path length will then be $\pi\left(\frac{60+110}{2}\right)=267\,\text{mm}.$

Flux Density

The requirements of phase displacement and angle error limit the working flux density of the core. An estimate of flux density can be made by considering one working condition, preferably one likely to be most stringent. So, considering phase displacement at the 20% full load condition.

$I_1=30$ amps $\qquad \theta=45'$

From phase diagram, $\sin\theta=\frac{I_r}{I_1}$

$\therefore I_r=I_r \text{ Sin } \theta=30\times.013$

$=0.4$ A

$H_r=\frac{I_r}{L_m}=\frac{.04}{0267}$

$=1.5$ A/M

By inspection of resolved component curves for TS grade core material:

$-H_r=1.5$ when $B_m=60$mT.

If the flux density at 20% F. L condition is chosen at 60mT, it will rise to 300mT at full load add other points which can be checked for error. If for any condition the phase displacement is excessive, a lower flux density must be chosen.

Condition (%Full Load)	120%	100%	20%	10%
Primary current I_1 (amps)	180	150	30	15
B max (mT)	360	300	60	30
H_r(from curves) A/m	45	4.0	1.5	0.95
I_r(Hr × 0.267)	1.2	1.068	0.4	0.307
$\theta(\text{Sin}^{-1} I_r/I_1)$	23'	2.4'	45'	58.5'
H_w (from curves) Am	5.2	45	1.05	0.6
$I_w(H_w\times0.267)$	1.39	1.20	0.28	0.16

$E(I_w/I_1 \times 100)$ %	0.77	0.80	0.94	0.07
1 turn compensation %	-0.67	-0.67	-0.67	-0.67
Compensation error e_1 %	0.1	0.13	0.27	0.4

Compensation

Assuming the phase angle displacements are within allowable limits, ratio error is calculated for each condition as shown above, and a turns ratio correction is chosen which will make them acceptable. In this case, 1 turn correction is made by reducing the secondary winding to 149 turns.

Cross Sectional Area

By choosing the working flux density at full load the required cross sectional area is calculated thus:

Voltage across Burden at the full load = 2.5 volts.

Allowing secondary winding resistance 0.1 ohms.

Then additional voltage for the internal burden = 0.1 Volts.

Total secondary E.M.F. = 2.6 volts.

For 149 turn secondary,

$$\text{Volts / Turn} = \frac{2.6}{149} = 0.0175 \text{ Volts}$$

$$\text{At rated condition } B_m = 0.3 \text{ Tesla}$$

$$\text{By transformer equestion } \frac{V}{T} = .0222 \times B_m \times Afe$$

$$\therefore \quad \text{Net C.S.A Afe} = \frac{0.175}{0.222 \times 0.3} = 2.63 \text{cm}^2$$

Allowing 0.95 space factor, Gross C.S.A $= 2.77 \text{cm}^2$

Final Dimensions

Before fixing final dimensions, take account of possible core degradation during winding. If protected by a case, this will be small, but it is prudent to allow the 20% extra area for a core taped, wound and impregnated. In this example, a strip width of a 20 mm with a buildup of 17 mm gives a final core dimension of,

I.D. – 60 mm.

O.D. – 94 mm.

Length – 200 mm.

3.2 Electronic Instruments for Measuring Basic Parameters

Amplified DC Meters

The direct coupled amplifier suits well for the very low frequency applications. Since the circuit especially works at low frequencies, the coupling capacitor and transformers are not used because electrical size of these components become very large at these frequencies.

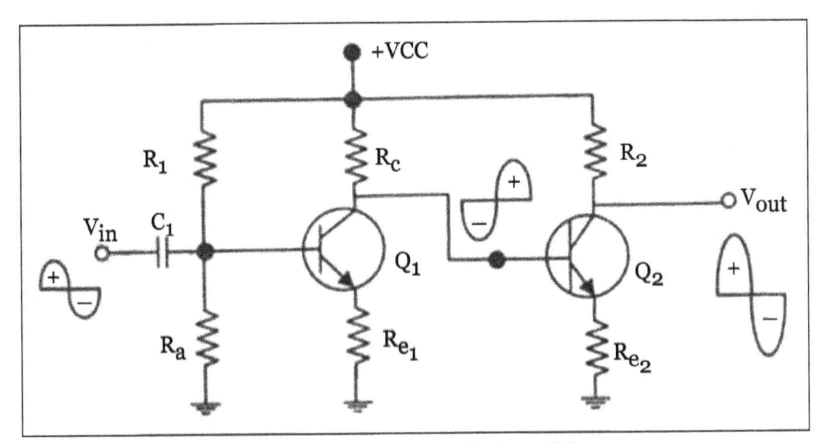

Three stage direct coupled amplifier.

The above diagram shows the circuit of three stage direct coupled amplifier. It uses complementary transistor. Thus, the first stage uses npn transistor, second stage uses pnp transistor and so on. In this circuit the output of first transistor is directly connected to input of second transistor.

Working: The weak signal is applied to the input of first transistor Q1. Due to the transistor action, an amplified output is obtained across the collector load RC of transistor Q1 .This voltage drives base of the second transistor and amplified output is obtained across its collector load. In this way, direct coupled amplifier raises strength of weak signal.

Advantages:

- The circuit arrangement is simple because of the minimum use of resistors.
- The circuit has low cost because of the absence of expensive coupling devices.

Disadvantages:

- It cannot be used for the amplification of high frequencies.

- The operating point is shifted due to the temperature variations.

Applications of Direct Coupled Amplifier

These are used to amplify low frequency (<10 Hz) signals. For example amplifying the photoelectric current, thermocouple current etc.

3.2.1 AC Voltmeters using Rectifiers

The PMMC used in the D.C. voltmeters can be effectively used in a.c. voltmeters. The rectifier is used to convert a.c. voltage to be measured, to d.c. This D.C., if needed is amplified and then given to the PMMC meter. The PMMC meter gives the deflection proportional to quantity to be measured. The r.m.s. value of an alternating quantity is given by the steady current (d.c.) that when flowing through a given circuit for a given time produces the same amount of heat as produced by alternating current which when flowing through the same circuit for the same time. The r.m.s value is calculated by measuring quantity at equal intervals for one complete cycle. Then squaring each quantity, average of squared values is obtained. The square root of this average value is the r.m.s. value. If the waveform is continuous then instead of squaring and calculating mean, the integration is used. Mathematically the r.m.s. value of the continuous AC voltage having time period T is given by:

$$V_{rms} = \sqrt{\frac{1}{T}\int_{0}^{T} V_{in}^{2}\, dt}$$

The 1/T term indicates mean value or average value.

For purely sinusoidal quantity:

$$V_{rms} = 0707\, V_m$$

Where,

V_m = peak value of the sinusoidal quantity.

If the a.c. quantity is continuous then average value can be expressed mathematically using an integration as:

$$V_{av} = \frac{2}{T}\int_{0}^{T/2} V_m\, dt$$

The interval T/2 indicated the average over half a cycle. For purely sinusoidal quantity:

$$V_{av} = \frac{2}{\pi} V_m = 0.636\, V_m$$

Where,

V_m = Peak value of the sinusoidal quantiry.

The form factor is the ratio of r.m.s. value to the average value of an alternating quantity.

$$K_t = \frac{\text{r.m.s value}}{\text{average value}} = \text{form factor}$$

Basic Rectifier Type Voltmeter

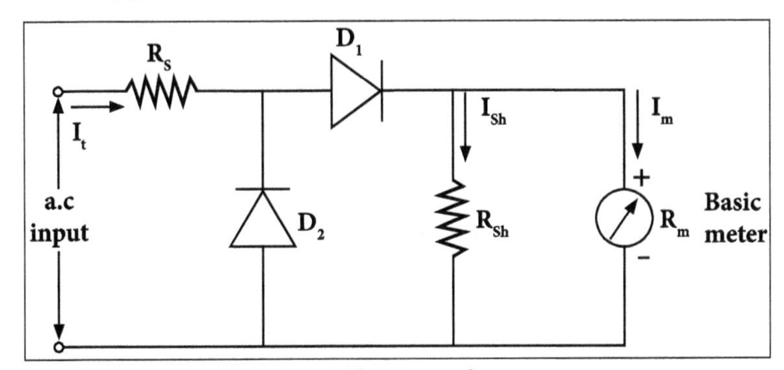

Basic rectifier type voltmeter.

The diodes D_1 and D_2 are used for rectifier circuit. The diodes show the nonlinear behaviour for the low currents hence to increase current through diode D_1, the meter is shunted with a resistance Rsh' this ensures high current through diode and its linear behavior. When a.c. input is applied, for the positive half cycle, the diode D_1 conducts and causes the meter deflection proportional to average value of that half cycle. In the negative cycle, the diode D_2 conducts and D_2 is reverse biased. The current through the meter is in opposite direction and hence meter movement is bypassed. Thus due to diodes, rectifying action produces pulsating D.C. and the meter indicates the average value of the input.

AC voltmeter using full wave rectifier:

The a.c. voltmeter using full wave rectifier is achieved by using the bridge rectifier consisting of four diodes, as shown in the Figure.

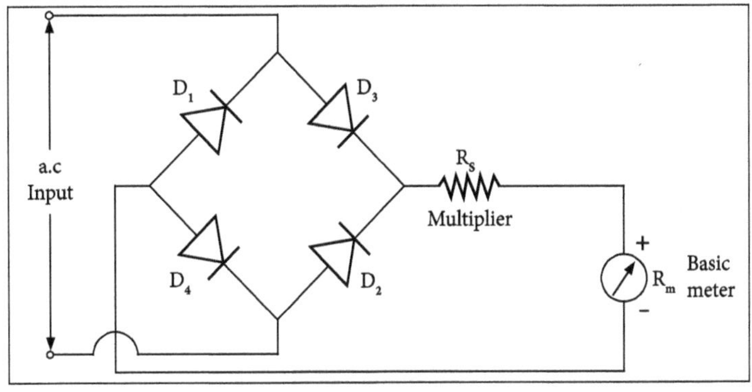

A.C voltmeter using full wave rectifier.

3.2.2 True RMS Voltmeter

A true-rms device (rms = root mean square) is one of three tools that can measure alternating current (ac) or ac voltage:

- True-rms digital multimeters (or clamp meter).
- Average-responding digital multimeter (or clamp meter).

Oscilloscope

Only the first two are commonly used, and both can accurately measure standard (pure ac) sinusoidal waveforms.

Yet true-rms meter is widely preferred because only it can accurately measure both sinusoidal and non-sinusoidal ac waveforms.

Sinusoidal (sine) waves: Pure, without distortion, with symmetrical transitions between peaks and valleys.

Non sinusoidal waves: Waves with distorted, irregular patterns—spikes, pulse trains, squares, triangles, sawtooths and any other ragged or angular waves.

rms = root mean square. rms essentially calculates equivalent direct current (dc) value of an ac waveform. More technically, it determines the "effective," or dc heating value, of any ac wave shape.

An average-responding meter uses averaging mathematical formulas to accurately measure the pure sinusoidal waves. It can measure non-sinusoidal waves, but with uncertain accuracy.

A sophisticated true-rms meter can accurately measure both pure waves and the more complex non-sinusoidal waves. Waveforms can be distorted by nonlinear loads such as variable speed drives or computers. An averaging meter attempting to measure distorted waves up to 40% low or 10% high in its calculations.

Multimeter type	Response to sine wave	Response to square wave	Response to single phase diode rectifier	Response to 3 ∅ diode rectifier
Average responding	Correct	10 % high	40 % low	5 % to 30 % low
True – rms	Correct	Correct	Correct	Correct

The need for true -rms meters has grown as the possibility of non-sinusoidal waves in circuits has greatly increased in recent years. Some examples:

- Variable-speed motor drives.
- Electronic ballasts.
- Computers.
- HVAC.

Solid-State Environments

In these environments, current occurs in short pulses rather than smooth sine wave drawn by a standard induction motor. The current wave shape can have a dramatic effect on a current clamp reading.

In addition, a true-rms meter is the better choice for taking measurements on power lines where ac characteristics are unknown.

3.2.3 Considerations for choosing an Analog Voltmeter

In choosing an analog voltmeter the following factors are to be considered.

- Input Impedance: The input impedance or resistance of the voltmeter should be as high as possible. It should always be higher than the impedance of the circuit under measurement to avoid the loading effect. The shunt capacitance across the input terminals also determines the input impedance of the voltmeter. At higher frequencies the loading effect of the meter is noticeable, since the shunt capacitance reactance falls and the input shunt reduces the input impedance.
- Voltage Ranges: The voltage ranges on the meter scale may be in a 1-3-10 sequence with 10 db separation or a 1.5-5-15 sequence or in a single scale calibrated in decibels. In any case, the scale division should be compatible with the accuracy of the instrument.
- Decibels: For measurements covering a wide range of voltages, the use of the decibel scale can be very effective, e.g., in the frequency response curve of an amplifier, where the output voltage is measured as a function of the frequency of the applied input voltage.
- Sensitivity v/s Bandwidth: Noise consists of unwanted frequencies. Since noise is a function of the bandwidth, a voltmeter with a narrow bandwidth picks up less noise than a large bandwidth voltmeter. In general, an instrument with a bandwidth of 10 Hz-10 MHz has a sensitivity of 1 mV. Some voltmeters whose bandwidth extends up to 5 MHz may have a sensitivity of 100 μ.V.

- Battery Operation: A voltmeter (VTVM) powered by an internal battery is essential for field work.
- AC Current Measurements: Current measurements can be made by a sensitive ac voltmeter and a series resistor. To summarise, the general guidelines are as follows.
    - For dc measurement, select the meter with the widest capability meeting the requirements of the circuit.
    - For ac measurements involving sine waves with less than 10% distortion, the average responding voltmeter is most sensitive and provides the best accuracy.
    - For high frequency measurement (> 10 MHz), the peak responding voltmeter with a diode probe input is best. Peak responding circuits are acceptable if inaccuracies caused by distortion in the input waveform are allowed (tolerated).
    - For measurements where it is important to find the effective power of waveforms that depart from the true sinusoidal form, the rms responding voltmeter is the appropriate choice.

3.2.4 Digital Voltmeters (Block Diagrams only)

Voltmeter is an electrical measuring instrument which is used to measure potential difference between two points. The voltage to be measured may be AC or DC. Two types of voltmeters i.e., analog and digital are available for the purpose of voltage measurement. Analog voltmeters generally contains a dial with a needle moving over it according to the measured value and hence displaying the value of the same.

With the passage of time, analog voltmeters are replaced by digital voltmeters due to the same advantages associated with the digital systems. Although the analog voltmeters are not fully replaced by digital voltmeters, still there are many places where the analog voltmeters are preferred over digital voltmeters.

Digital voltmeters display the value of AC or DC voltage being measured directly as the discrete numerical instead of a pointer deflection on the continuous scale as in the analog instruments.

Working Principle

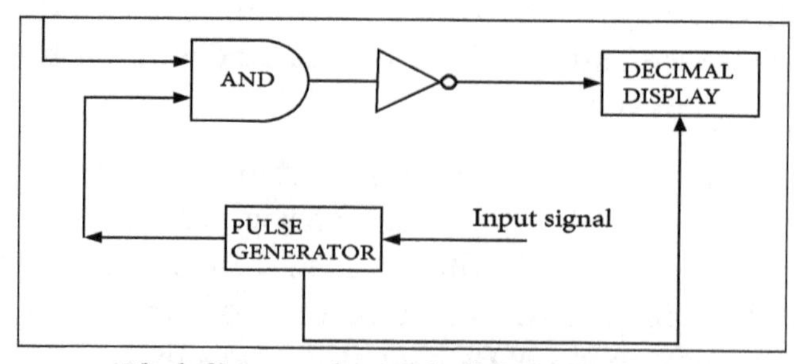

Block diagram of a simple digital voltmeter.

- Input signal: It is basically a signal i.e., voltage to be measured.
- Pulse generator: Actually, it is a voltage source. It uses the digital, analog or both the techniques to generate a rectangular pulse. The width and frequency of the rectangular pulse is controlled by the digital circuitry inside generator while amplitude and rise & fall time is controlled by analog circuitry.
- AND gate: It gives high output only when both the inputs are high. When a train pulse is fed to it along with rectangular pulse, it provides an output having train pulses with duration same as that of the rectangular pulse from the pulse generator.

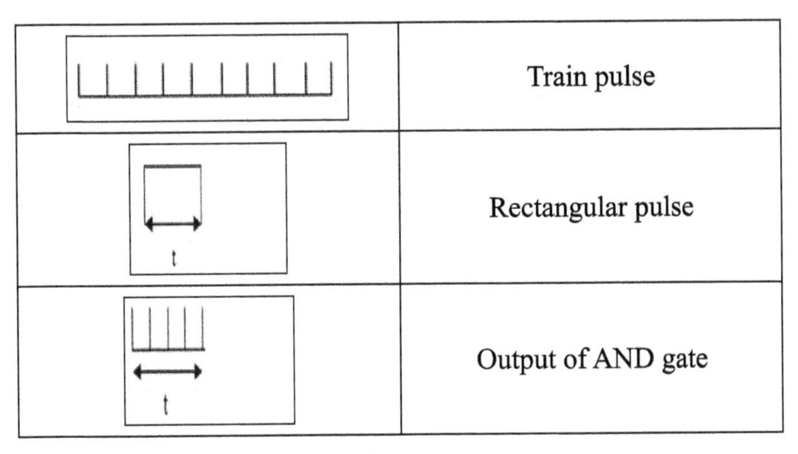

- NOT gate: It inverts the output of AND gate.

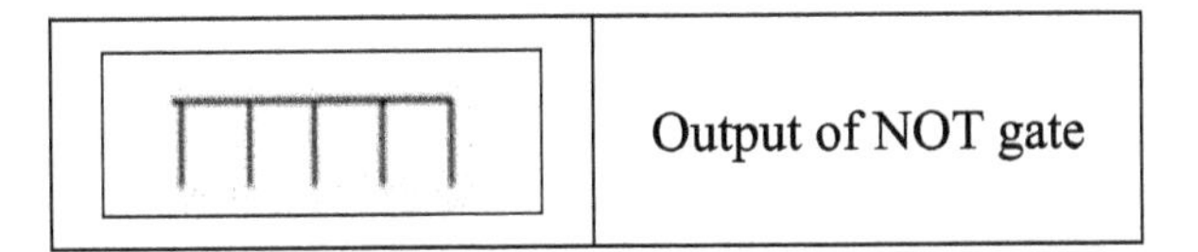

Decimal Display

It counts the numbers of impulses and thus the duration and display value of voltage on LED or LCD display after calibrating it.

The working of a digital voltmeter as follows:

- Unknown voltage signal is fed to pulse generator which generates a pulse whose width is proportional to the input signal.
- Output of pulse generator is fed to one leg of AND gate.
- Input signal to the other leg of the AND gate is a train of pulses.
- Output of the AND gate is positive triggered train of duration same as the width of the pulse generated by the pulse generator.
- This positive triggered train is fed to the inverter which converts the positive triggered train into a negative triggered train.

- Output of the inverter is fed to the counter which counts the number of triggers in the duration which is proportional to input signal i.e. voltage under measurement.

Hence, counter can be calibrated to indicate the voltage in volts directly.

We can see the working of digital voltmeter that it is nothing but an analog to digital converter which converts an analog signal to a train of pulses, the number of which is proportional to the input signal. So a digital voltmeter can also be made by using any one of the A/D conversion methods.

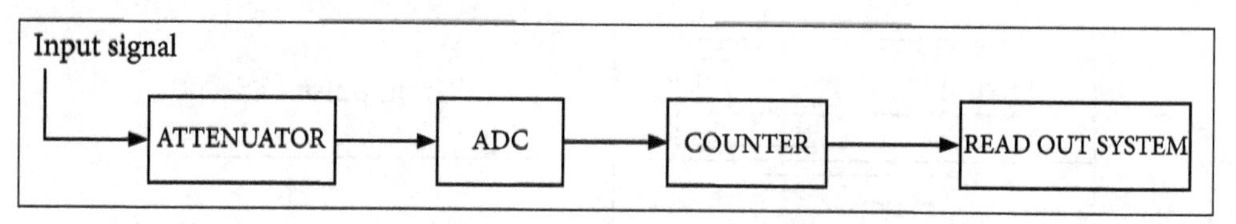

On the basis of A/D conversion method used digital voltmeters are classified as:

- Integrating type voltmeter.
- Ramp type digital voltmeter.
- Successive approximation type digital voltmeter.
- Potentiometric type digital voltmeters.
- Continuous balance type digital voltmeter.

Digital voltmeters are also replaced by digital millimeters due to its multitasking feature i.e., it can be used for measuring the current, voltage and resistance. But still there are some fields where separate digital voltmeters are being used.

Advantages of Digital Voltmeters

- Errors on account of parallax and approximation is entirely eliminated.
- Read out of DVMs is easy as it eliminates the observational errors in the measurement committed by the operators.
- Output can be fed to the memory devices for storage and future computations.
- Reading can be taken very fast.
- Compact and cheap.
- Versatile and accurate.
- Portability increased.
- Low power requirements.

3.2.5 Q-meter

The ratio of inductive reactance to the effective resistance of the coil is called the quality factor or Q factor of the coil.

So,

$$Q = X_L / R = \omega L / R$$

A high value of Q is always desirable as it means high inductive reactance and low resistance. The low value of Q indicates that the resistance component is relatively high and so there is a comparatively large loss of power.

The effective resistance of the coil differs from its DC resistance because of eddy current and skin effects and it varies in a highly complex manner with the frequency.

For this same reason, the Q value is rarely calculated by the determination of R and L.

One possible way for determination of Q is by using the inductance bridge but such bridge circuits are rarely capable of giving the most accurate measurements, when the value of Q is high. So special meters are used for determination of Q accurately.

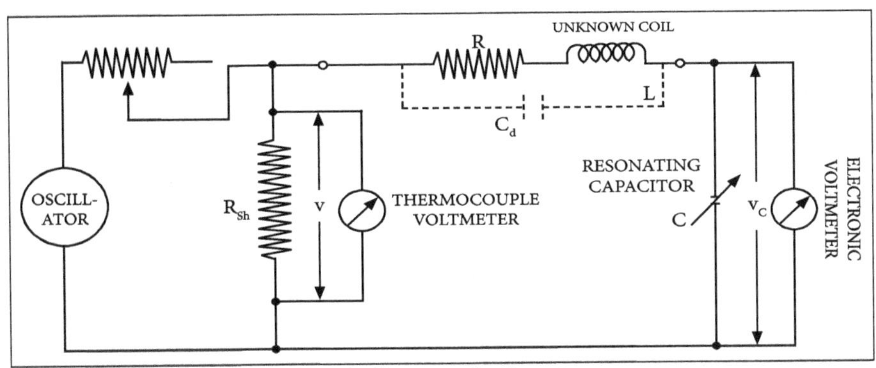

Circuit diagram of a Q-meter.

Q-meter is an instrument which is designed for the measurement of Q-factor of coil as well as for the measurement of electrical properties of coils and capacitors.

This instrument operates on the principle of the series resonance i.e. at resonate condition of an AC series circuit, the value of voltage across the capacitor is equal to applied voltage times Q of the circuit.

If the voltage applied across the circuit is kept at constant, then the voltmeter connected across the capacitor can be calibrated to indicate the Q value directly.

Circuit diagram of a Q-meter is shown is figure. A wide-range oscillator with frequency range from 50 kHz to 50 MHz is used as a power supply to the circuit.

The output of the oscillator is shorted by a low-value resistance, R_{sh} which is usually of the order of 0.02 ohm. So it introduces almost no resistance into the oscillatory circuit and represents a voltage source with a very small or of almost negligible internal resistance.

The voltage across the shunt resistance R_{sh}, V is measured by a thermocouple meter and the voltage across the capacitor, V_c is measured by an electronic voltmeter.

For carrying out the measurement, the value of unknown coil is connected to the test terminals of the instrument and the circuit is tuned to resonance either by varying the frequency of the oscillator or by varying the value of resonating capacitor C.

Readings of voltages across capacitor C and shunt resistance R_{sh} are obtained and Q-factor of the coil is determined as follows:

By definition Q-factor of the coil is given by:

$$Q = X_L / R$$

When the circuit is under resonance condition it can be written as:

$$X_L = X_C$$

$$\text{Or} \quad IX_L = IX_C = V_C$$

Voltage applied to the circuit is given by:

$$V = IR$$

$$\text{So, } Q = X_L / R = IX_L / R = V_C / V$$

The actual Q-factor of the coil is greater than the calculated Q-factor.

This difference is usually very small and can be neglected except when the resistance of the coil under test is relatively very small compared to the shunt resistance R_{sh}.

The inductance of the coil is computed from the known values of the frequency f and the resonating capacitor C as follows:

At resonance:

$$X_L = X_C$$

Or,

$$2\Pi f_L = 1/2\Pi f_C$$

Or,

$$L = 1/(2\Pi f)^2 \text{ Henry}$$

3.3 Oscilloscope

The basic concept of the oscilloscope is that it displays waveforms in a two dimensional format. The vertical axis is normally used to plot incoming voltage, and the horizontal axis is normally used as a time axis. In this way waveform voltage can be displayed as a function of time.

Oscilloscope Probes

The oscilloscope, or scope for short, is essential tool for fault finding for electronics development, repair or diagnostics work. The oscilloscope enables waveforms on various parts of the circuit to be viewed in a graphical format. To enable the oscilloscope to connect to required points, oscilloscope probes or scope probes are required.

Although it is possible to use a signal line and earth return connection to form a simple oscilloscope probe, this approach does not provide optimum performance as both the electrical and mechanical aspects need to be considered to meet the necessary requirements.

A whole variety of scope probes can be bought and used. Fortunately, it is essential to know which types to use, and what scope probe specifications may be chosen.

Types:

- Passive oscilloscope probes.
- Active oscilloscope probes.

Types of Oscilloscope

Since the first oscilloscopes were developed and produced different types of oscilloscope have been developed. These have arrived as the technology has developed and they have been focused on the particular applications and measurements. As a result, when choosing an oscilloscope, one of first decisions to be made is regarding the basic form of the oscilloscope that is needed.

- Analogue oscilloscope,
- Analogue storage oscilloscope,
- Digital oscilloscope,
- Digital storage oscilloscope,
- Digital phosphor oscilloscope,
- Digital sampling oscilloscope.

Oscilloscopes have been in use in the electronics industry for many years. Even with many new developments taking place and a shift to greater levels of software within products there is no lessening of the importance of the oscilloscopes. As technology develops, so technology used within scopes has enabled them to provide higher levels of performance and provide new and useful functions. With continuing movement of technology, oscilloscope technology will also move forwards.

Cathode Ray Tube (CRT)

The CRT is the heart of the CRO. The CRT generates the electron beam, accelerates the beam, deflects the beam and has a screen where beam becomes visible as a spot.

The main parts of the CRT are:

- Electron gun,
- Deflection system,
- Fluorescent screen,
- Glass tube of Envelope,
- Base.

Electron Gun

The electron gun section of the CRT provides a sharply focused electron beam directed towards the fluorescent coated screen. This screen starts from the thermally heated cathode, emitting the electrons. The control grid is given negative potential with respect to cathode. This grid controls number of electrons in the beam, going to the screen.

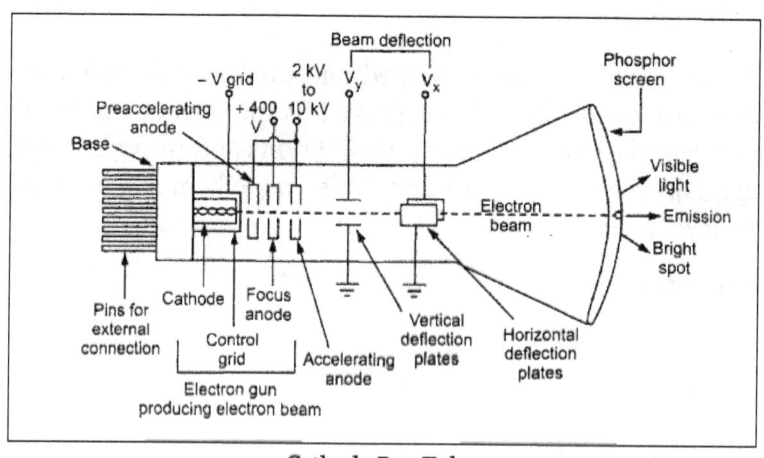

Cathode Ray Tube

The momentum of the electrons determines intensity or brightness of the light emitted from the fluorescent screen due to electron bombardment. The light emitted is usually of the green colour. Because the electrons are negatively charged, a repulsive force is created by applying negative voltage to the control grid. This negative control voltage can be made variable.

Since the electron beam consists of many electrons, the beam tends to diverge. This is because the similar charges on the electron repel each other. To compensate for such repulsion forces, an adjustable electrostatic field is created between two cylindrical anodes, which result into the required acceleration of the electrons.

Both focusing and the accelerating anodes are cylindrical in shape having small openings located in the centre of each electrode, coaxial with the tube axis. The pre-accelerating and accelerating anodes are connected to a common positive high voltage which varies between 2 kV to 10 kV. The focusing anode is connected to a lower positive voltage of about 400 V to 500 V.

Delay Line

As the name suggests that, this circuit is used to, delay the signal for a period of time in the vertical section of CRT. The input signal is not applied directly to the vertical plates because the part of the signal gets lost, when the delay Time not used. Therefore, input signal is delayed by a period of time.

Deflection System

When the electron beam is accelerated it passes through the deflection system, with which beam can be positioned anywhere on the screen.

The deflection system of the CRT consists of two pairs of parallel plates, referred to as the vertical and horizontal deflection plates. One of the plates in each set is connected to ground (0 V). To the other plate of each set, the external deflection voltage is applied through an internal adjustable gain amplifier stage. To apply deflection voltage externally, an external terminal, called the Y input or the X-input, is available.

As shown in the figure, the electron beam passes through these plates. A positive voltage applied to the Y input terminal (V_Y) causes beam to deflect vertically upward due to the attraction forces, while a negative voltage applied to Y input terminal will cause the electron beam to deflect vertically downward, due to the repulsion forces.

Also, a positive voltage applied to X-input terminal (V_x) will cause the electron beam to deflect horizontally towards the right, while a negative voltage applied to X-input terminal will cause the electron beam to deflect horizontally towards left of the screen. The amount of vertical or horizontal deflection is directly proportional to the corresponding applied voltage.

When the voltage are applied simultaneously to vertical and horizontal deflecting plates, electron beam is deflected due to the resultant of these two voltages. The face of the screen can be considered as an x – y plane. The (x, y) position of beam spot is thus directly influenced by the horizontal and the vertical voltages applied to deflection plates V_x and V_Y, respectively.

The horizontal deflection (X) produced will be proportional to horizontal deflecting voltage, V_x, applied to X-input

$\therefore \quad XV_x \; X = K_x \; V_x.$

Where K_x is constant of proportionality.

The deflection produced is usually measured in cm or number of divisions, on the scale in the horizontal direction.

Then $K_x = X/V_x$ where K_x expressed as cm/volt or division/volt, is called horizontal sensitivity of the oscilloscope.

Similarly, vertical deflection (Y) produced will be proportional to the vertical deflecting voltage, VY, applied to the Y-input.

$\therefore$ Y V_Y and K_Y, vertical sensitivity will be expressed as cm/volt or division/volt.

The values of vertical and horizontal sensitivities are selectable and adjustable through multi positional switches on the front panel that controls the gain of corresponding internal amplifier stage. The bright spot of the electron beam can thus trace (or plot) x-y relationship between the two voltages, V_x and V_y.

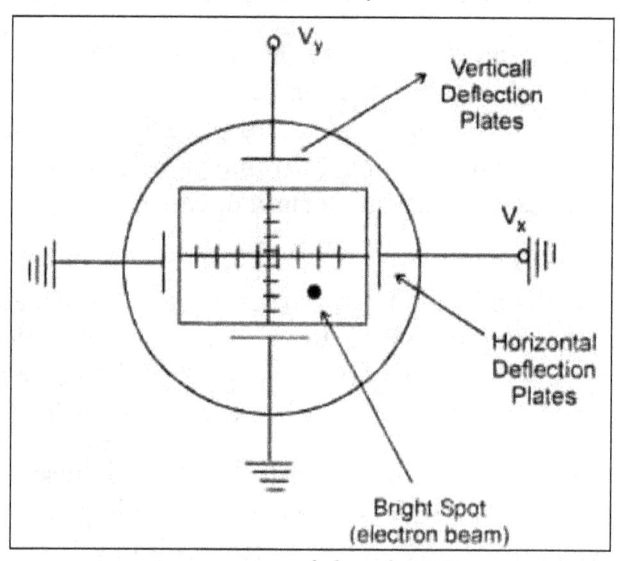

Arrangement of plates in CRT.

Fluorescent Screen

The light produced by screen does not disappear immediately when bombardment by electrons ceases, i.e., when the signal becomes zero. The time period for which trace remains on the screen after the signal becomes zero is known as persistence. The persistence may be as short as a few micro second, or as long as tens of seconds or even minutes.

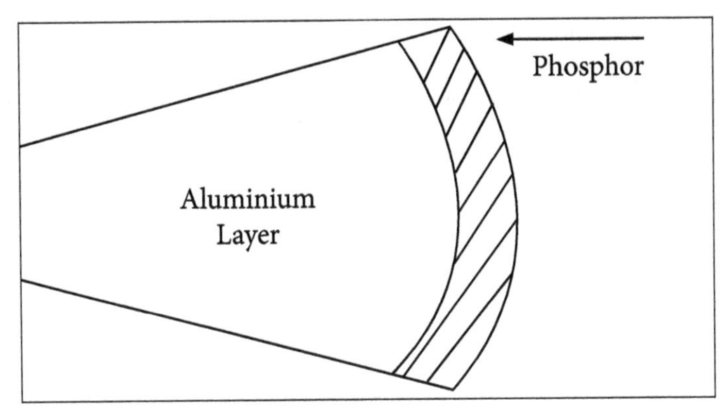

Fluorescent Screen.

Medium persistence traces are mostly used for general purpose applications. Long persistence traces are used in the study of transients. Long persistence helps in study of transients since the trace is still seen on the screen after the transient has disappeared. Short persistence is needed for extremely high speed phenomena. The screen is coated with a fluorescent material called phosphor which emits light when bombarded by electrons. There are various phosphors available which differ in colour, persistence and efficiency.

One of the common phosphor is willemite, which is zinc, orthosilicate, $ZnO + SiO_2$, with traces of manganese. This produces the familiar greenish trace. Other useful screen materials include compounds of cadmium, zinc, magnesium and silicon.

The kinetic energy of the electron beam is converted into both light and the heat energy when it hits the screen. The heat so produced gives rise to "phosphor burn" which is damaging and sometimes destructive. This degrades the light output of phosphor and sometimes may cause complete phosphor destruction. Thus phosphor must have high resistance to avoid accidental damage.

Similarly the phosphor screen is provided with an aluminium layer called aluminizing The cathode ray tube. Many phosphor materials having different excitation times and colours as well as different phosphorescence times are available.

Glass Tube: All components of a CRT are enclosed in an evacuated glass tube called envelope. This allows the emitted electrons to move about freely from one end of tube to the other end.

Base: The base is provided to the CRT through which the connections are made to various parts.

Deflection Defocussing and its Causes: Whenever an electron beam is deflected from the axial direction, the spot on the fluorescent screen tends to distort and enlarge. This phenomenon due to which spot does not remain in focus and get distorted is called deflection defocusing.

The various reasons of defocusing are:

- The distance of various points on the screen from the electron gun is not same. The distance from electron gun to screen is greater at the edges due to which defocusing results.
- The non-uniformity in the electric and magnetic deflection fields is used for the deflection. Due to this part of beam passing through stronger fields get more deflected and part passing through weaker field gets less deflected which results in defocussing
- All the electrons in a beam cannot have exactly same velocity. So due to unequal velocities of the electrons in the beam, defocusing results.

Digital CRO: The digital storage oscilloscope replaces the unreliable storage methods used in analog storage scopes with the digital storage with the help of memory. The memory can store data as long as required without degradation.

In this, the waveform to be stored is digitized and then stored in a digital memory. The power to the memory is small and hence stored image can be displayed indefinitely. Once the waveform is digitized then it can be further loaded into the computer and can be analysed.

Block Diagram: The block diagram of digital storage oscilloscope is shown in the figure. In all the oscilloscopes, the input signal is applied to the amplifier and attenuators section. The oscilloscope uses same type of amplifier and attenuators-circuitry as used in the conventional oscilloscopes. The attenuated signal is then applied to the vertical amplifier.

The vertical input after passing through vertical amplifier is digitized by an analog to digital converter to create a data set which is stored in the memory. The data set is processed by the microprocessor and then sent to the display.

To digitize the analog signal, A/D converter is used. The output of vertical amplifier is applied to the A/D converter section. The main requirement of A/D converter in the digital storage oscilloscope is its speed, while in digital voltmeters accuracy and resolution were the main requirements. The digitized output needed only in binary form and not in BCD. The successive approximation type of A/D converter is most often used in the digital storage oscilloscope.

The digitizing of the analog input signal means taking samples at periodic intervals of the input signal. The rate of sampling should be at least twice as fast as the highest frequency according to sampling theorem. This ensures no loss of information. The sampling rates as high as 100,000 samples per second is used. This requires very fast conversion rate of A/D converter.

If a 12-bit converter is used, 0.025% resolution is obtained while if 10- bit A/D converter is used then resolution of 0.1 % is obtained. Similarly with 10-bit A/D converter, the frequency response of 25 kHz is obtained. The total digital memory storage capacity is 4096 for a single channel, 2048 for two channels each and 1024 for four channels each. The sampling rate and memory size are selected depending on the rate and memory size are selected depending upon the duration and the waveform to be recorded.

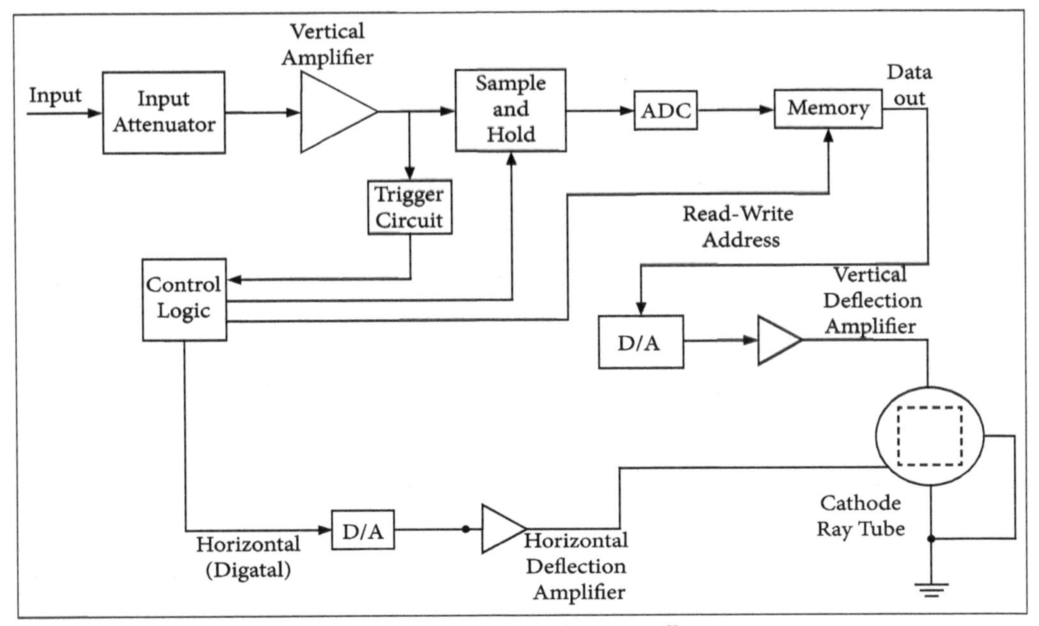

Block diagram of digital storage oscilloscope.

Once the input signal is sampled, the A/D converter digitizes it. The signal is then captured in the memory. Once it is stored in memory, many manipulations are possible as memory can be read out without being erased.

The digital storage oscilloscope has three modes. They are:

- Roll mode - Very fast varying signals are displayed clearly in this mode.
- Store mode - This is called refresh. In this the input initiates trigger circuits. Memory is refreshed for each triggers.
- Hold or same modes - This is automatic refresh mode. By pressing hold a save button, overwriting can be stopped and previously saved signal gets locked.

Advantages of D.S.O:

- Easier to operate and has more capability.
- The storage time is infinite.
- The display flexibility is available.
- The cursor measurement is possible.
- The characters can be displayed on screen along with the waveform.

- The X-Y plots, B-H curve, PV diagrams can be displayed.
- The pre-trigger viewing feature allows to display the waveform before trigger pulse.
- Keeping the records is possible by transmitting the data to computer system where the further processing is possible.
- Signal processing is possible which includes translating the raw data into finished information.

Analogue Storage Scope

It is sometimes necessary to be able to display a signal for a period of time. One situation where this may be required is for signals that have a very long period and the normal persistence of a display would mean that the trace would decay before the whole waveform was complete. A storage facility could be required for single shot applications where the single trace would need to be displayed over a period of time to examine the trace.

For these and many other situations, it is necessary to have a storage facility on scope where it can display the trace for longer than would normally be possible.

Analogue storage scopes use a special cathode ray tube with long persistence facility. A special tube is used with an arrangement to store charge in the area of display where the electron beam had struck, thereby enabling the fluorescence to remain for much longer than attainable on normal displays.

These cathode ray tubes had the facility to vary the persistence, although if very bright traces were held over the long periods of time, they would have the possibility of permanently burning trace onto the screen. Accordingly these storage displays should be used with care.

3.3.1 Measurement of Frequency

The frequency of a signal is measured using oscilloscope in two methods. They are:

- Using calibrated oscilloscope.
- Using uncalibrated oscilloscope.

Measurement of Frequency using Calibrated Oscilloscope

It is the indirect method of measurement of frequency. In this method, frequency of the unknown signal is measured by measuring its time period.

Initially, the unknown frequency signal is applied to vertical inputs of the CRO. Now the horizontal sweep is turned ON and the display appearing on the screen is adjusted by varying the different control knobs provided on the front panel of CRO, till the signal is displayed on the screen. After obtaining the display of good deflection, count number of horizontal division for a complete cycle. From counted horizontal divisions, the time period is computed as:

$$T = m * n$$

Where,

m=Number of division in one complete cycle.

n=Setting of time base =Time/Division.

From the measured time period of the signal, unknown frequency is calculated as:

$$f = 1/T$$

3.3.2 Phase Angle and Time Delay using Oscilloscope

A common time measurement performed with oscilloscope is a measurement of phase angle between two waveforms. The traces on the oscilloscope must be properly set up. Then the phase angle is determined using information from the display.

Setting up the waveform display:

- Begin by setting vertical amplifier coupling switch (AC-GND-DC) to AC. This will block the DC on the signal, making it easier to center the trace.
- Adjust the vertical amplifier VOLTS/DIV control to get the largest possible display on screen. It may be necessary to adjust the vertical amplifier POSITION control to center the trace as we do this. Be sure the red CAL knob is in detent position. This must be done for both channels.
- Adjust horizontal time base SEC/DIV to get about one cycle of the waveform on the display. We must have at least one complete cycle, but a few cycles to maximize accuracy.
- Set the vertical amplifier coupling switch (AC-GND-DC) to GND on each channel. This will result in a horizontal line for each channel.

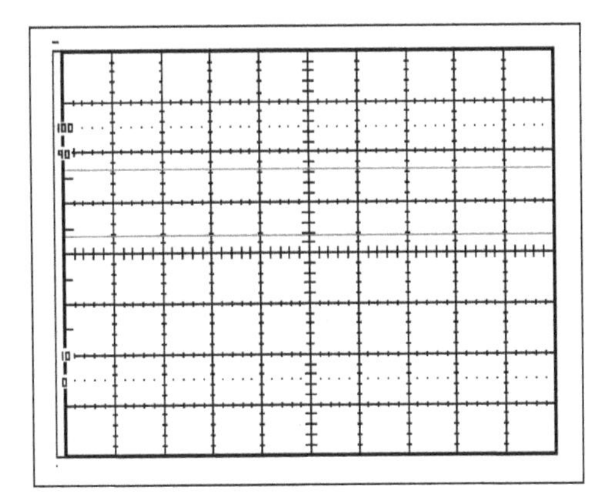

- Using the vertical position control for each channel, position the two lines on center graticule line.

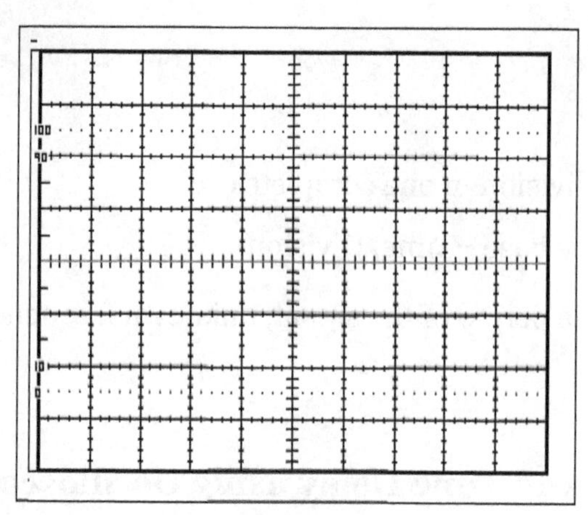

- Switch both vertical amplifier coupling switches (AC-GND-DC) to AC. This results in the both traces being seen.

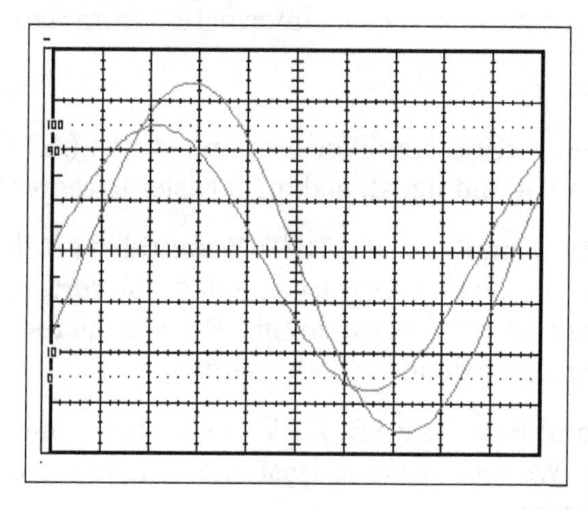

- Use the horizontal position control to position one of the traces at an intersection of graticule lines.

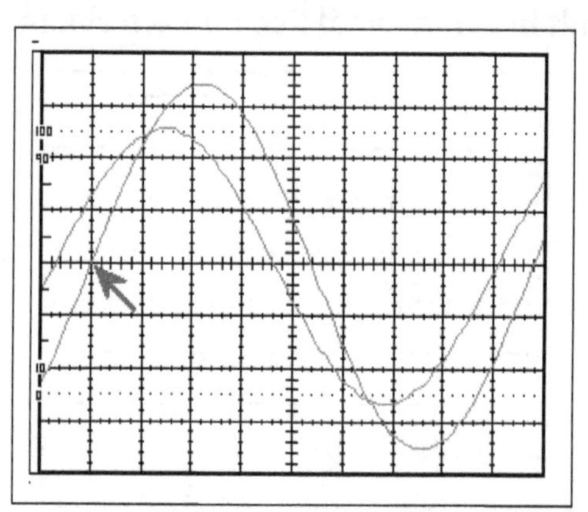

Measuring Phase

The formula used to calculate phase angle is:

$$\theta = \frac{t}{T} \times 360^{\circ}$$

- Count the number of horizontal divisions from where the trace crosses the intersection point to where the other waveform crosses center graticule line. This is t:

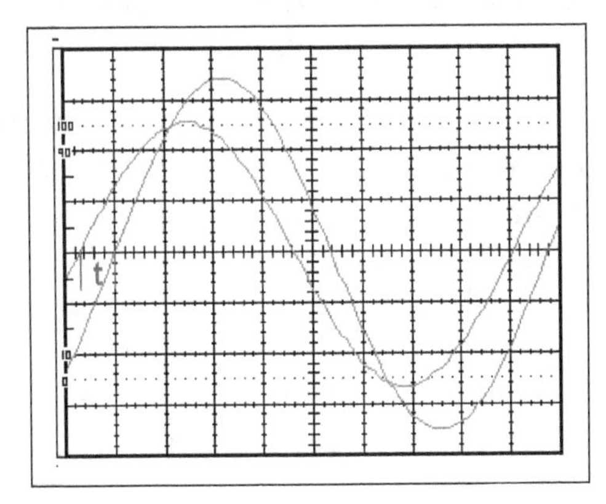

- Measure the period of one of the waves. This is T:

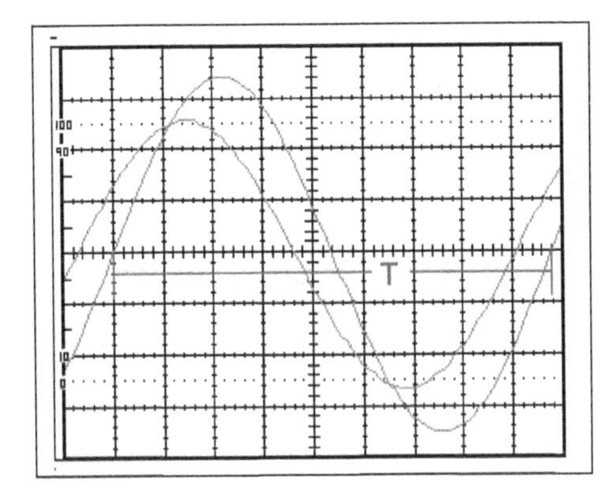

- The phase angle is:

$$\theta = \frac{t}{T} \times 360^{\circ}$$

Be sure to record which wave is leading.

Time Delay

A uniform time delay implies that we could remove delay from phase response to achieve a horizontal line at 0 degrees (which means no delay). If we can't get to a straight horizontal line by removing the delay or if the horizontal line we then achieve is at some angle other than 0 degrees, the system being measured will exhibit waveform distortion. So it seems that a way to determine whether a delay is constant over the band of frequencies is to work backwards and see whether removing some constant delay will get us to zero degrees everywhere within that band.

To remove delay at a given frequency, take the time delay being tried (in seconds) and multiplied by 360 times the frequency. Do this for each data frequency and add the result to each raw phase angle (in degrees), remembering that you can also add or subtract 360 degrees as many times as necessary at each point to reduce answer to within +/-180 degrees. Do this for various delay values until a best approximation of a straight line is reached.

This is not very practical to do by hand. It is much more practical to simply apply a known complex waveform to system and see whether it passes through uncorrupted in the time domain. That is the motivation for square wave or triangle wave testing. However, such simple time domain testing may tell you that there is waveform distortion, but would not clearly indicate whether any problem uncovered is due to frequency response magnitude or phase error. In addition, meaningful square wave tests are not easily conducted on loudspeakers because of inability to remove the effects of room reflections or echoes which also strongly alter the shapes of the reproduced periodic waveforms.

With IMP or LAUD generated quasi-anechoic responses, we can easily remove delay mathematically using the computer until best approximation to a straight line is achieved on the plot over the frequencies of interest. This can be done by trial and error without much effort. If the line is essentially near 0 degrees (if at 180 degrees, we can reverse the speaker leads), the delay of the speaker is uniform and complex waves made up from these frequencies will pass phase aligned. The magnitude of frequency response must also be flat to achieve waveform integrity. In IMP or LAUD, fixed amounts of delay can be subtracted from or added to a phase plot by using the [F9] key.

We can get a close starting value for amount of delay to be removed by measuring the distance from speaker to the mike and multiplying the value in feet by 0.886 (or the value in meters by 2.91). This result is the number of milliseconds it took for the signal to reach mike after it left the speaker. For this to be valid (in the case of IMP and LAUD), measurement should be made with a "Cal" normalization from signal at the crossover input (in dual channel mode for LAUD). We should also set the first time marker to a placement of "1" before any transformations so that the time window includes the entire time of flight. (For further information about normalization, time markers, windows, and measurement devices, see the series of IMP articles in the Speaker Builder issues 1,2,3,4 and 6 of 1993, the IMP Guide, or the Liberty Audiosuite manual).

We should again mention that we probably need not have such extremely uniform phase response or waveform shape integrity for the good sound reproduction. It is nothing unusual to find that the phase curve of a very good speaker straightens only over the short portions of the audio band and often away from 0 degrees, revealing non-uniform delay.

3.4 Counters and Analyzers

Frequency Selective Wave Analyzer

The wave analyzer consists of a very narrow pass-band filter section which can be tuned to a particular frequency within the audible frequency range (20 Hz – 20 kHz). The block diagram of a wave analyzer is as shown in Figure below.

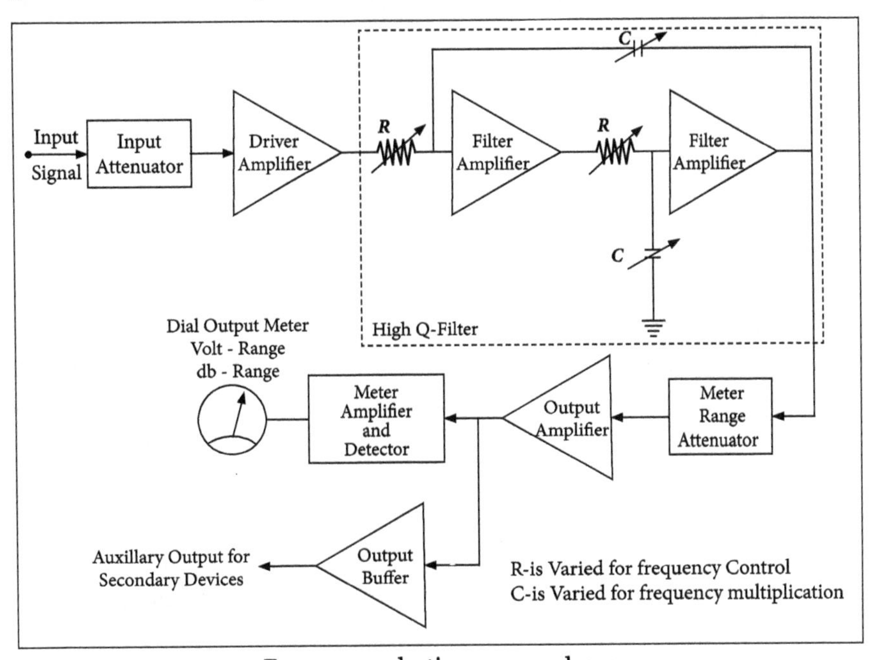

Frequency selective wave analyzer.

A basic wave analyzer is shown in Figure (a). It consists of a primary detector, which is a simple LC circuit. This LC circuit is adjusted for resonance at the frequency of the particular harmonic component to be measured.

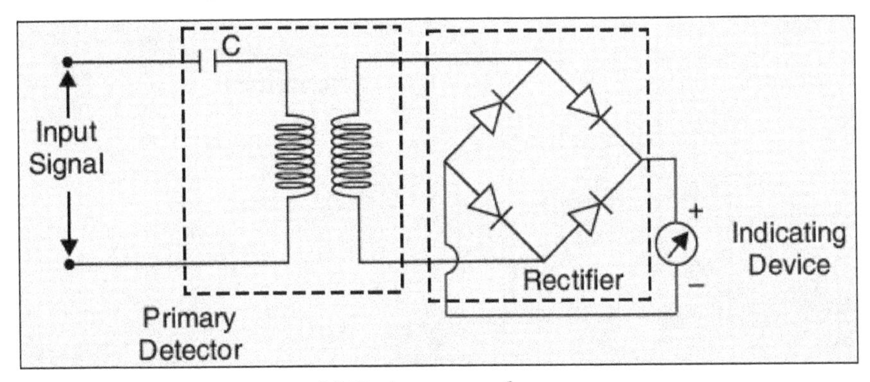

(a) Basic wave analyzer.

The intermediate stage is a full wave rectifier, to obtain the average value of the input signal. The indicating device is a simple dc voltmeter that is calibrated to read the peak value of the sinusoidal input voltage. Since the LC circuit is tuned to a single frequency, it passes only the frequency to which it is tuned and rejects all other frequencies. A number of tuned filters, connected to the indicating device through the selector switch, would be required for a useful Wave analyzer.

The complex wave to be analyzed is passed through an adjustable attenuator that serves as a range multiplier and permits a large range of signal amplitudes to be analyzed without loading amplifier.

The output of the attenuator is then fed to a selective amplifier, which amplifies selected frequency. The driver amplifier applies the attenuated input signal to a high-Q active filter. This high-Q filter is a low pass filter that allows the frequency which is selected to pass and reject all others. The magnitude of this selected frequency is indicated by the meter and the filter section identifies the frequency of the component. The filter circuit consists of a cascaded RC resonant circuit and amplifiers. For selecting the frequency range, the capacitors used are of the closed tolerance polystyrene type and the resistances used are precision potentiometers. The capacitors are used for range changing and the potentiometer is used to change the frequency within selected pass-band, hence this wave analyzer is also called a Frequency selective voltmeter.

The entire AF range is covered in decade steps by switching capacitors in the RC section.

The selected signal output from final amplifier stage is applied to the meter circuit and to an unturned buffer amplifier. The main function of the buffer amplifier is to drive output devices, such as recorders or electronics counters. The meter has several voltage ranges as well as the decibel scales marked on it. It is driven by an average reading rectifier type detector. The wave analyzer must have extremely low input distortion, undetectable by the analyzer itself. The band-width of the instrument is very narrow, typically about 1% of selective band given by the following response characteristics shown in Figure.

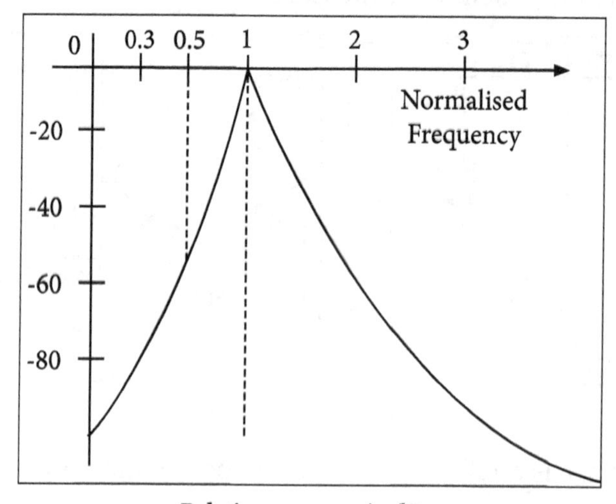

Relative response in dBs.

Heterodyne Wave Analyzer

Wave analyzers are useful for measurement in the audio frequency range only. For measurements in the RF range and above (MHz range), an ordinary wave analyzer cannot be used. Hence, special types of wave analyzers working on the principle of heterodyning (mixing) are used. These wave analyzers are known as heterodyne wave analyzers.

In this wave analyzer, the input signal to be analyzed is heterodyned with the signal from the internal tunable local oscillator in the mixer stage to produce a higher IF frequency.

By tuning the local oscillator frequency, various signal frequency components can be shifted within the pass-band of the IF amplifier. The output of the IF amplifier is rectified and applied to the meter circuit.

An instrument that involves the principle of heterodyning is the Heterodyning tuned voltmeter, shown in Figure.

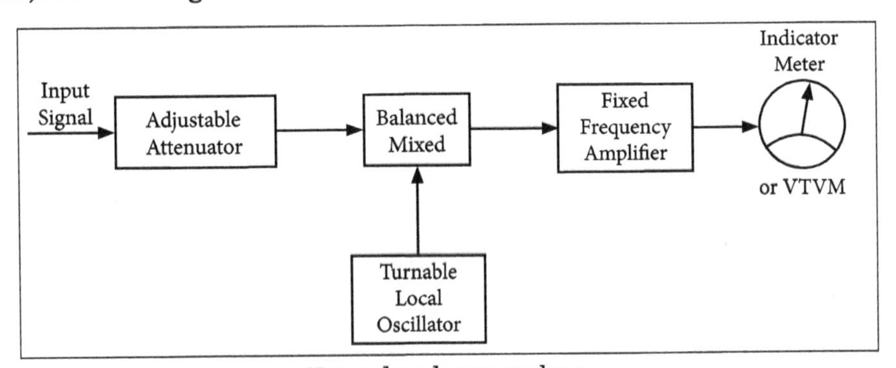

Heterodyned wave analyzer.

The input signal is heterodyned to the known IF by means of a tunable local oscillator. The amplitude of the unknown component is indicated by the VTVM or output meter. The VTVM is calibrated by means of signals of known amplitude.

The frequency of the component is identified by the local oscillator frequency, i.e. the local oscillator frequency is varied so that all the components can be identified. The local oscillator can also be calibrated using input signals of known frequency. The fixed frequency amplifier is a multistage amplifier which can be designed conveniently because of its frequency characteristics. This analyzer has good frequency resolution and can measure the entire AF frequency range. With the use of a suitable attenuator, a wide range of voltage amplitudes can be covered. Their disadvantage is the occurrence of spurious cross-modulation products, setting a lower limit to the amplitude that can be measured.

Two types of selective amplifiers find use in heterodyne wave analyzers. The first type employs a crystal filter, typically having a centre frequency of 50 kHz. By employing two crystals in a band-pass arrangement, it is possible to obtain a relatively flat pass-band over a

4 cycle range. Another type uses a resonant circuit in which the effective Q has been made high and is controlled by negative feedback. The resultant signal is passed through a highly selective 3-section quartz crystal filter and its amplitude measured on a Q-meter.

When knowledge of the individual amplitudes of the component frequency is desired, a heterodyne wave analyzer is used. A modified heterodyne wave analyzer is shown in Figure. In this analyzer, the attenuator provides the required input signal for heterodyning in the first mixer stage, with the signal from a local oscillator having a frequency of 30 - 48 MHz.

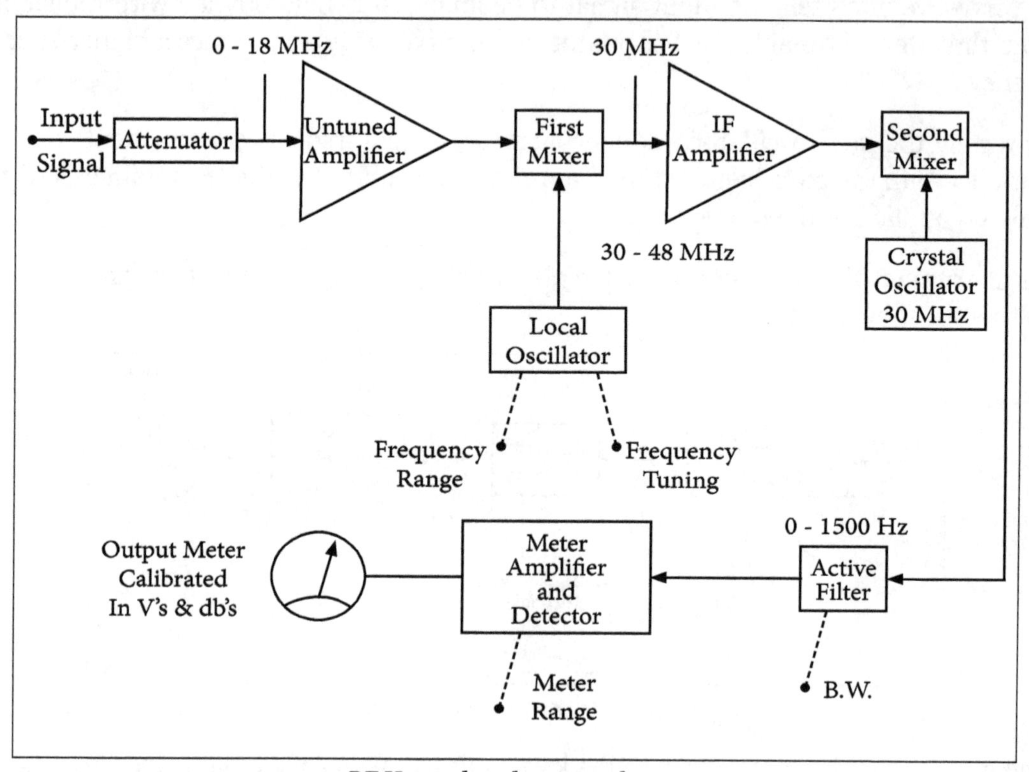

RF Heterodyned wave analyzer.

The first mixer stage produces an output which is the difference of the local oscillator frequency and the input signal, to produce an IF signal 30MHz.

This IF frequency is uniformly amplified by the IF amplifier. This amplified IF signal is fed to the second mixer stage, where it is again heterodyned to produce a difference frequency or IF of zero frequency.

The selected component is then passed to the meter amplifier and detector circuit through an active filter having a controlled band-width. The meter detector output can then be read off on a db-calibrated scale, or may be applied to a secondary device such as a recorder.

This wave analyzer is operated in the RF range of 10 kHz - 18 MHz, with 18 over-lapping bands selected by the frequency range control of the local oscillator. The bandwidth, which is controlled by the active filter, can be selected at 200 Hz, 1 kHz and 3 kHz.

Harmonic Distortion

Distortion analyzer measures the total harmonic power present in the test wave rather than the distortion caused by each component. The simplest method is to suppress the fundamental frequency by means of a high pass filter whose cut off frequency is a little above the fundamental frequency. This high pass allows only the harmonics to pass and the total harmonic distortion can then be measured.

Types of harmonic distortion analyzers based on fundamental suppression are as follows:

- Employing a Resonance Bridge.
- Wien's Bridge Method.
- Bridged T-Network Method.

Network Analyzer

Network analyzer is used to measure reflection coefficients, transmission coefficients, insertion loss, return loss, S parameters and more. It takes care of the measurement of components, devices and sub modules.

Spectrum Analyzer

Spectrum analyzer is used to measure the carrier power harmonics, spurious, side-bands, level, phase noise and more.

Network analyzer is mainly designed to take care of measurements related to phase such as group delay which is not possible with the spectrum analyzer.

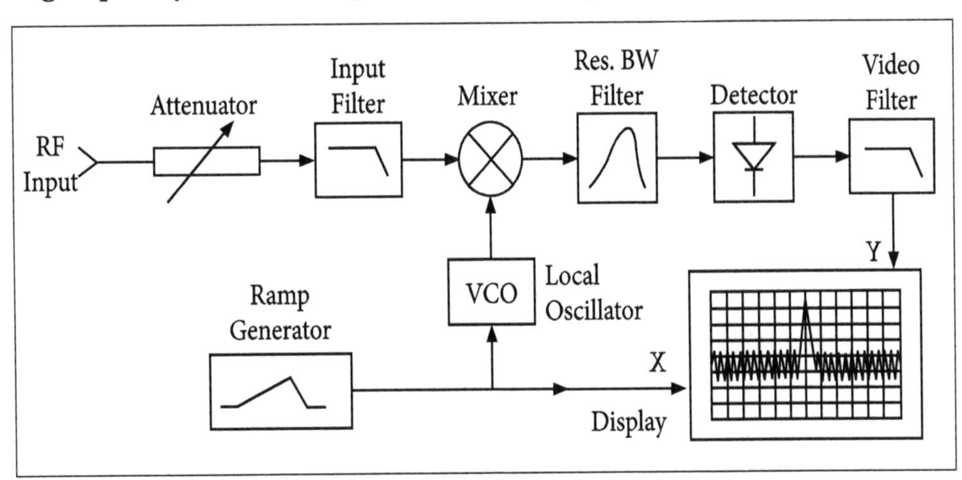

Block diagram of spectrum analyser.

Spectrum Analyzer helps discover unwanted signals and network analyzer helps to measure known signals.

Spectrum Analyzer is just a receiver which displays the signal fed to its RF input port from any RF transmitting device through a cable or with antenna.

Modern real time spectrum analyzers even decode and display complex broadband RF signals emitted/acquired using various wireless devices which for example include WLAN, WI MAX, GSM, LTE, Zigbee and more. To do so additional base band circuitry is incorporated in the spectrum analyzer.

Network analyzer consists of the source and multiple receivers and measures broadband frequency signal using techniques such as power and the frequency sweep.

The spectrum analyzers usually will have wider range IF bandwidths compared to most of network analyzers.

Network analyzer provides higher measurement accuracy compared to spectrum analyzer due to vector error correction feature.

While doing measurement with spectrum analyzer, it is easy to place a marker on display, but interpretation of results are difficult.

While doing measurement with network analyzer, it is hard to place a marker on display, but interpretation of results are very easy.

Frequency Counter

A frequency counter is an electronic instrument, or component, that is used for measuring frequency. Frequency counters usually measure number of oscillations or pulses per second in a repetitive electronic signal. Such an instrument is sometimes referred to as a cymometer.

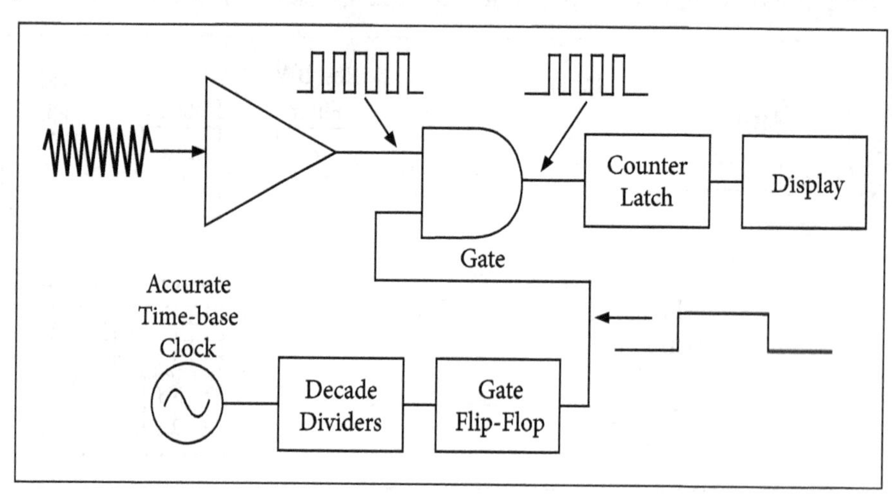

Permissions

All chapters in this book are published with permission under the Creative Commons Attribution Share Alike License or equivalent. Every chapter published in this book has been scrutinized by our experts. Their significance has been extensively debated. The topics covered herein carry significant information for a comprehensive understanding. They may even be implemented as practical applications or may be referred to as a beginning point for further studies.

We would like to thank the editorial team for lending their expertise to make the book truly unique. They have played a crucial role in the development of this book. Without their invaluable contributions this book wouldn't have been possible. They have made vital efforts to compile up to date information on the varied aspects of this subject to make this book a valuable addition to the collection of many professionals and students.

This book was conceptualized with the vision of imparting up-to-date and integrated information in this field. To ensure the same, a matchless editorial board was set up. Every individual on the board went through rigorous rounds of assessment to prove their worth. After which they invested a large part of their time researching and compiling the most relevant data for our readers.

The editorial board has been involved in producing this book since its inception. They have spent rigorous hours researching and exploring the diverse topics which have resulted in the successful publishing of this book. They have passed on their knowledge of decades through this book. To expedite this challenging task, the publisher supported the team at every step. A small team of assistant editors was also appointed to further simplify the editing procedure and attain best results for the readers.

Apart from the editorial board, the designing team has also invested a significant amount of their time in understanding the subject and creating the most relevant covers. They scrutinized every image to scout for the most suitable representation of the subject and create an appropriate cover for the book.

The publishing team has been an ardent support to the editorial, designing and production team. Their endless efforts to recruit the best for this project, has resulted in the accomplishment of this book. They are a veteran in the field of academics and their pool of knowledge is as vast as their experience in printing. Their expertise and guidance has proved useful at every step. Their uncompromising quality standards have made this book an exceptional effort. Their encouragement from time to time has been an inspiration for everyone.

The publisher and the editorial board hope that this book will prove to be a valuable piece of knowledge for students, practitioners and scholars across the globe.

Index

Printed in the USA
CPSIA information can be obtained
at www.ICGtesting.com
JSHW051621061123
51533JS00005B/49